Code of Practice for Programme Management in the Built Environment

Code of Practice for Programme Management in the Built Environment

Second Edition

CIOB

THE CHARTERED INSTITUTE OF BUILDING

WILEY Blackwell

Registered Offices
John Wiley & Sons, Inc., 111 River Street, Hoboken, NJ 07030, USA
John Wiley & Sons Ltd, The Atrium, Southern Gate, Chichester, West Sussex, PO19 8SQ, UK

For details of our global editorial offices, customer services, and more information about Wiley products visit us at www.wiley.com.

Wiley also publishes its books in a variety of electronic formats and by print-on-demand. Some content that appears in standard print versions of this book may not be available in other formats.

Library of Congress Cataloging-in-Publication Data

Names: Chartered Institute of Building (Great Britain), author.
Title: Code of practice for programme management in the built environment /
     The Chartered Institute of Building.
Description: Second edition. | Hoboken, NJ : Wiley-Blackwell, 2024. |
     Includes bibliographical references and index.
Identifiers: LCCN 2024003438 (print) | LCCN 2024003439 (ebook) | ISBN
     9781394192434 (paperback) | ISBN 9781394192458 (adobe pdf) | ISBN
     9781394192441 (epub)
Subjects: LCSH: Building–Superintendence. | Project management.
Classification: LCC TH438 .C626 2024 (print) | LCC TH438 (ebook) | DDC
     658.4/04 – dc23/eng/20240212
LC record available at https://lccn.loc.gov/2024003438
LC ebook record available at https://lccn.loc.gov/2024003439

Cover Design: Wiley
Cover Images: © Ashley Cooper/Getty Images; pyramid/line graphic - created by Wiley - concept provided by author

Set in 10/13pt Franklin Gothic by Straive, Chennai, India
Printed and bound by CPI Group (UK) Ltd, Croydon, CR0 4YY

C9781394192434_080324

Contents

CHAPTER 1

Programme Management in Context 1

5

Stage D: Implementation 73

Stage E: Benefits Realisation and Transition 95

Templates 131

Key Roles: Skills and Competencies   147

Foreword

The Code of Practice for Programme Management was first published in 2016 and, in that relatively short time, we have seen many challenging changes in the way that built environment projects are delivered.

This second edition is a natural development from the first edition and builds on the sixth edition of the Code of Practice for Project Management published in 2022, authored by two experts with UK and international expertise in this field.

The first edition of this Code of Practice defined a programme as:

> *A collective of related projects is coordinated to achieve desired benefits more effectively than when managing them as a group of individual projects.*

This definition has been retained for the second edition.

The term 'Built Environment' has been retained as we continue to see projects that are not solely construction or development-related. Some of the client sectors, for example, highways, rail, airports, shipping or nuclear are likely to incorporate projects that are not related to construction. Accordingly, I believe that this code has relevance to a very wide reader base, which goes beyond construction.

Six core themes have been identified as key to the life cycle strategy of the programme. These are 1. developing capability, 2. procuring value, 3. digitalisation and visualisation, 4. cultural identity and ethics, 5. sustainability and its governance and 6. platform thinking.

The need for programme management arises when benefits obtained in a coordinated manner are greater than the sum of individual project benefits obtained in isolation.

This need has continued to grow due to several factors, including being seen as a way to help transform the industry as a whole, which has continued to struggle with productivity, collaboration and innovation. In addition, it is seen as an opportunity to manage the availability of resources and skills in the industry and obtain wider beneficial societal change beyond solely the strategic objectives of the client organisation.

As this second edition was being prepared, society was experiencing major disruptions, including the pandemic, high inflation and a war on the borders of Europe. These global issues are influencing the energy supply. This is at the same time as environmental challenges and an ongoing emphasis on social value and sustainability.

All this points to a current and future operating environment that is more uncertain than we have experienced in recent times.

These changes provide a greater emphasis on the application of programme management becoming more dynamic, which requires actively organising and managing.

I believe the application of this Code of Practice can assist those responsible for programmes in this rapidly changing world. It will also serve as an important textbook reference in academia.

I thank all those involved in the production of this code who have offered advice and guidance as this edition was being written. My special thanks to those who have spent time reviewing and commenting on the work of the authors. I also thank colleagues at CIOB for their work in coordinating the numerous parts of the code.

Mike Foy OBE FCIOB MBA FCMI
CIOB President 2021/2022

Acknowledgements

I am delighted that the Chartered Institute of Building is publishing this updated version of the Code of Practice for Programme Management. As Past President Mike Foy mentions in his foreword, the built environment and construction sector are ever-changing, and there is perhaps more need than ever to look to the role of Programme Manager, with their overview of work and the opportunity to anticipate potential changes.

The previous edition of this code of practice was ably steered by CIOB Fellows and other industry experts, who focused on producing a practical document for a discipline that is present in many different sectors. CIOB is a broad church, covering an impressive range of built environment sector roles and, in turn, this publication is intended for an audience of programme management professionals who are active across a range of industries.

The process to update this second edition of our Code of Practice for Programme Management reflects both the depth of expertise of the profession and the breadth of sectors it spans, with contributions from built environment specialists and other professional institutions providing their insights and expertise.

A list of participants and the organisations represented is included in this book. I want to take this opportunity to thank all the CIOB members who contributed their time in producing this revised Code of Practice for Programme Management. I also want to acknowledge all those who have supported our publications over the years – our community of members who contribute to our publications has always been generous in sharing their time and expertise.

Caroline Gumble
Chief Executive
Chartered Institute of Building

List of Figures

Working Group (WG) of the Code of Practice for Programme Management

Chair

Michael J Foy OBE, FCIOB, MBA, FCMI   Working Group Chair, CIOB President 2021/2022

Technical Authors

Dr. Tahir Hanif FCIOB, FAPM, FACostE, FIC FRICS

Principal PMO and Project Controls Consultant – Costain; Honorary Professor - University of the West of Scotland

Dr. Simon Addyman MSc, PhD, FAPM   Associate Professor of Project Management, UCL

Working Group

Gildas André   MBA, MSc, BSc (Hons), MAPM, MCIOB, Director GAN Advisory Services

David Haimes MCIOB   David Haimes Consulting Limited

Dr. Ruth Murray-Webster HonFAPM   Director, Potentiality UK

Miss Niki French FICE   Head of Utilities, HS2

Andy Stanley FCIOB, FAPM, AMICE, NECReg

Programme Management Practitioner.

Dr. Paul Sayer   Publisher: John Wiley & Sons Limited

Dr. Ashwini Konanahalli BArch, MSc, PhD, FHEA

Reader in Construction Management, University of the West of Scotland

The following also contributed in the development of the Code of Practice for Programme Management.

David Philp MSc, BSc, FCIOB FICE FRICS FCInstCES

Chief Value Officer, Cohesive Group

Jamie Strathearn MRICS MCIOB   Head of Programme, Cost, Procurement, Pre-Construction - Marks and Spencer

Francis Ho FCIOB   Partner at Charles Russell Speechlys LLP

Milan Radosavljevic PhD UDIG MIZS-CEng

Vice-Principal and Pro Vice-Chancellor (Research, Innovation and Engagement)

Mr. Wes Beaumont BSc, MSc, MCIOB   Associate Vice-President, Digital Transformation Leader, AECOM US

Neil Thompson FIET, MCIOB, MDAMA   Director, AtkinsRealis

Bill McElroy FAPM, FICE   Independent Consultant

Glossary

Benefits	A (directly or indirectly) measurable improvement resulting from an outcome perceived as an advantage by one or more stakeholders and that contributes towards one or more organisational strategic objective(s).
Benefits management	The identification, definition, monitoring, realisation and optimisation of benefits within and beyond a programme.
Benefits profile	Used to define each benefit (and dis-benefit) and provide a detailed understanding of what will be involved and how the benefit will be realised.
Benefits realisation manager (BRM)	Supports programme manager by taking the responsibility for benefits identification, mapping and realisation – ensures that necessary business benefits are realised.
Benefits realisation plan	Used to monitor the realisation of benefits across the programme and set governing controls.
Business change manager (BCM)	Responsible for ensuring that the objectives have been sufficiently and accurately defined, managing the transition activities and undertaking and determining whether the intended benefits have been realised.
Business partner	Organisations that have a business or financial interest in the outcome of the programme.
Clients	Persons using the services of a professional entity or those who are procuring products or services from a professional entity. In legal context, a client may instruct a professional entity to act on the client's behalf. In the programme sense, this document defines clients as 'the body or group that procures the services of professionals to initiate and deliver projects or a programme of projects'.
Customer	Persons who are paying for a product or a service but not necessarily in the legal context represented by the professional entity.
Deliverable	What is to be provided as a result of an initiative or project – typically tangible and measurable.
Disbenefit	A (directly or indirectly) measurable decline resulting from an outcome perceived as negative by one or more stakeholders that may or may not affect one or more organisational strategic objective(s).
Issue	A relevant event that has happened or is likely to happen was not planned and requires management action.
Opportunity	A relevant but uncertain event that can have a favourable impact on objectives or benefits.

Outcome	The result of a change. Outcomes are desired when a change is conceived and are achieved as a result of the activities undertaken to reflect the change.
Output	The tangible or intangible effect of a planned activity or initiative.
Portfolio	A portfolio is a total collection of programmes and/or stand-alone projects managed by an organisation to achieve strategic objectives.
Programme	A programme is a collective of related projects coordinated to achieve desired benefits not possible from managing them as a group of individual projects.
Programme brief	Used to assess whether the programme is viable and achievable.
Programme communication manager	Supports the programme manager by managing all internal and external communication channels, developing the programme communications plan and ensuring governance of internal and external communication protocols.
Programme delivery plan (PDP)	A detailed description of what the programme will deliver, how and when it will be achieved and the financial implications of its delivery and implementation.
Programme financial manager	Deals with complex financial issues including funding arrangements, cash flow and financial governance. Responsible for programme financial plan, budget and financial reporting.
Programme financial plan	A financial statement that collects all the costs that have been identified in relation to implementing the programme – often the funding streams are also identified in this document.
Programme management board	A group established to support a programme sponsor in delivering a programme.
Programme management office (PMO)	The function providing information and governance for a programme and its delivery objectives – it can provide support to more than one programme.
Programme manager	The role responsible for the setup, management and delivery of a programme – typically allocated to a single individual; for large and complex programmes an organisation can be given this role.
Programme mandate	Expansion of the vision statement setting out in greater detail what it is that the programme needs to achieve in terms of the outcomes and what it is that the programme seeks to deliver.
Programme monitor	In certain privately funded programmes, a programme monitor (sometimes known as funder/lender/investor's advisor or monitor) may be appointed, on behalf of the funding entities, to safeguard the interests of the funders.
Programme sponsor	The main driving force behind a programme, appointed by the client and the point of accountability for the programme.
Programme sponsor's board	The driving group behind the programme, which provides the investment decisions and senior-level governance for the rationale and objectives of the programme.
Programme timescale plan	An overall delivery time schedule for the programme.
Project	A project is a temporary and transient undertaking created to achieve agreed objectives and produce and deliver a product, service or result
Risk	An uncertain event or set of events that, if it occurs, has an effect on the achievement of the objectives. A risk is measured by a combination of the probability of a perceived threat or opportunity occurring and the magnitude of its impact on objectives.

Stakeholder	Any individual, group or organisation that can affect, be affected by or perceive itself to be affected by a programme.
Transition	The changes that need to take place in business as usual, which are aimed to be managed, as project outputs are exploited in order to achieve programme outcomes.
Transition plan	The schedule of activities to cover the transition phase of the benefits realisation plan.
Value	The total sum of benefits to be derived from the programme less the total costs expended across the life of the business case.
Vision	A view of a better future that will be delivered by the programme.
Vision statement	A business vision for change setting out the intent and the benefits sought.

1 Programme Management in Context

1.1 Introduction

This second edition of the Code of Practice for Programme Management in the Built Environment is a natural development from the first edition and builds on the recently published sixth edition of the highly successful Code of Practice for Project Management for the Built Environment.

The first edition of this Code of Practice defined a programme as:

> a collective of related projects coordinated to achieve desired benefits more effectively than when managing them as a group of individual projects.

In this second edition, we retain this definition. In some organisations, programmes are created with a single business case aligned to a set of benefits. While for others, business cases and benefits are defined at project and not programme level. However, in both cases, the projects are delivered as a programme in order to achieve an organisation's strategic objectives more effectively.

We have retained 'for the built environment' because we continue to see many projects that are not solely construction or development-related. For example, if we consider some of the client sectors involved in creating new facilities and/or infrastructure, such as highways, rail, airports, shipping, nuclear, etc., all of these are likely to incorporate projects that are not related to construction. These may include disciplines such as information technology, human resources management (HRM), capacity building, marketing, etc. Indeed, even mainstream construction developments may include similar disciplines as self-contained projects within a programme.

The sixth edition of Project Management continues to provide the relevant guidance and procedural requirements for the successful management of individual projects. This second edition of Programme Management further develops the elements of functionality and procedures specific to the management and successful delivery of a number of related projects within the built environment, focusing at the programme level on the coordinated creation of value, which presents itself as a trade-off in the relationship between (see Figure 1.1):

(i) Benefits and outcomes (via outputs);

(ii) Risks and opportunities;

(iii) Requirements and objectives (incl. cost and time).

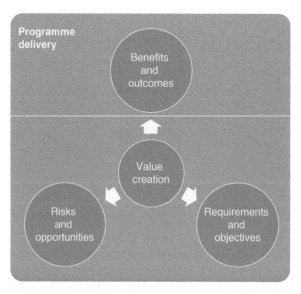

Figure 1.1 Programme Delivery Triangle.

1.1.1 Need for programme management in the built environment

When an organisation manages multiple related projects independently, issues such as a lack of willingness to share resources and knowledge or the use of different tools and techniques can limit the ability to monitor and measure performance across the related projects. This gives rise to problems of coordination and control, which will likely lead to outputs, outcomes and benefits becoming misaligned with an organisation's strategic objectives, resulting in the loss of value.

The need for programme management arises when benefits obtained in a coordinated manner are greater than the sum of individual project benefits obtained in isolation. Whereas traditionally a project is measured by the criteria of time, cost and quality, programmes are determined and measured by the strategic objectives and the benefits to the client organisation, which might otherwise not have been realisable had the projects been managed independently. Furthermore, programmes offer the ability to manage resources, risks and opportunities more efficiently across multiple projects. In this sense, programme management is a systematic approach to project grouping for the purpose of achieving benefits without leaving the process to chance.

The need for programme management in the built environment has continued to grow due to a number of factors, including:

(i) Programmes are being seen as an opportunity to better manage the availability of resources and skills in the industry;

(ii) There is a growing need for clients and suppliers to integrate government policy into the design and execution of their projects;

(iii) Programmes are being seen as a way to help transform the industry as a whole, which has continued to struggle with productivity, collaboration and innovation;

(iv) Programmes are being seen as an opportunity to bring about wider beneficial societal change beyond solely the strategic objectives of the client organisation.

In this context, the construction sector is shifting its focus from products to services, where programmes and their associated projects do not just deliver buildings; they deliver outputs, outcomes and benefits that create the additional value their stakeholders are seeking. This in turn gives firms in the sector a greater opportunity for survival in what is becoming an ever more competitive environment.

The 2012 London Olympics represents a good example of a programme where individual venues, infrastructure, legacy, etc., each represented separate sets of related projects all under the umbrella of the Olympic Delivery Authority (ODA). The ODA was tasked to act as the programme client, with a programme delivery partner, to achieve the desired benefits (including a long-term legacy) that would not have been achieved had they been managed individually. Another example is the UK's High Speed 2 rail infrastructure programme, where, for example, training academies are created to support the development of skills in the sector.

Clients are now seeing a programme of related projects as posing less risk and greater opportunity to achieve strategic objectives than a number of individual projects can when considered alone. Given the scope for variations in terminology and approaches that are possible within the practices of programme management, this Code of Practice establishes a clear and consistent understanding of the processes involved in managing programmes in the built environment, regardless of their size, nature or complexity.

1.1.2 Future programme management in the built environment

As we prepare and publish this second edition society is still emerging from the pandemic, experiencing a bifurcation of global supply chains, high inflation and facing a war on the borders of Europe. These are global issues that are influencing disruptions in energy supply and contributing to a period of economic uncertainty. This is in addition to the environmental challenge, continued developments in the digitalisation of society, and the ongoing emphasis on social value and sustainability through corporate social responsibility (CSR) practices and environment, social and governance (ESG) metrics.

All of these factors are having an influence on the transformation of our industry, from government policy to firm strategies, in ways that point to a current and future state that is more uncertain and in greater flux than we have experienced in recent history. When we look at recent studies in industry transformation,[1] what we see is that top-down approaches, such as government policy and firm strategies, are not enough to achieve the transformation that is needed for a more sustainable future. What is emerging is a greater acknowledgement that the practices and routines that practitioners engage in when delivering projects and programmes can have a significant positive and/or negative influence on industry transformation from within the programme or project itself.

In light of this, we see the practice of programme management in the built environment becoming ever more dynamic in nature. This dynamism needs actively organising and managing through the application of this Code of Practice. More specifically, the organisational routines engaged in by the multiple organisations involved in developing and delivering programmes need to be purposefully reproduced every time. This is because we now understand these routines as being the key mechanisms through which an organisation becomes capable of delivering its objectives. As programmes are temporary organisational arrangements (albeit often of a long duration), these routines can only be reproduced once the programme is underway and must continue to be reproduced as the programme moves through its life cycle.

[1] Glass, J., Bygballe, L.E., and Hall, D., 2022. Transforming construction: the multi-scale challenges of changing and innovating in construction. *Construction Management and Economics*, 40(11–12), pp. 855–864.

Programme management therefore is no longer simply about adapting a client organisation's capabilities from steady state A to steady state B, but one in which both current state A and future state B are uncertain and in flux. In this sense, programmes cannot be delivered solely through the mindless application of static models of practice, as these models in themselves cannot determine the programme outcomes.[2] Programme managers must move away from the assumption that once a particular programme arrangement has been established, it is already, once and for all, capable of delivering the client objectives. While this may seem self-evident, the stagnation of productivity in construction, the lack of investment in firm capabilities and rising concerns about worker welfare suggest otherwise.

This means that programme managers must continuously re-create capable programme organisations by adapting the guidance presented here to the specific situation and timing of both the individual programmes and their projects, as well as the context of the firms (clients, contractors, consultants, etc.) from which the programmes/projects are created.

This draws greater attention to how programme management practices operate at three specific interfaces:

(1)   between the client organisation and the programme;

(2)   between the supply chain firms and the programme;

(3)   between the programme and the projects (see Figure 1.2).

Each of the organisations involved will come to the programme with their own routines and capabilities. For programme managers, then, this becomes a matter of being able to organise for and manage the reproduction of a nexus of routines at these interfaces, as without doing so there is a risk of disorder that may lead to a misalignment with the client objectives. In the context we have described above, this is a challenge, yet it also offers an opportunity for programme management to act as an ideal organisational arrangement to deal with a multiplicity of competing factors and to transition society, the economy and firms to a more sustainable, value-driven future.

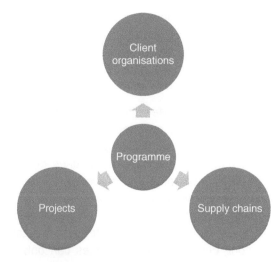

Figure 1.2 Organisational interfaces in the programme context. Adapted from Winch, G.M., Maytorena-Sanchez, E. and Sergeeva, N., 2022. Strategic project organizing. Oxford University Press.

---

[2]   Addyman, S., and Smyth, H., 2023. *Construction Project Organising*. John Wiley & Sons, Ltd.

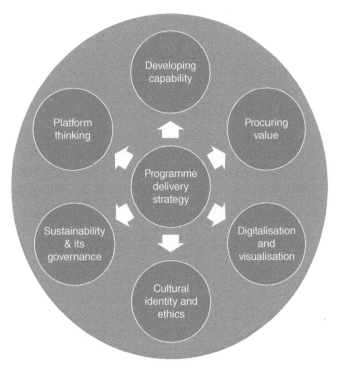

Figure 1.3 Six themes for programme management in the built environment.

In summary, then, programme management can act as an 'organisational bridge' between the longer-term strategic objectives of the firms (clients and suppliers), the capabilities they need to survive and prosper, and a more sustainable future for society more generally. In the following section, we look at six core themes that we have identified as being important for future programme management.

1.1.3 Core themes that must be integrated

In our review of current trends in the organisation and management of built environment programmes and projects, we have identified six themes that, in one form or another, need to be integrated into the life cycle strategy of the programme. We interpret these themes with due consideration of our definition of programme management presented above.

In addition to the guidance that we have taken from our working group for this second edition, we have drawn inspiration for these themes from work that has gone on over the last five years between government and firms to transform the industry. This includes, but is not limited to, outputs from the Transforming Construction Network+ research programme, The Construction Innovation Hub, The Construction Leadership Council, the Infrastructure and Project Authority Project Initiation Routemap and the industry professional bodies through updates in their bodies of knowledge and codes of practice, including the sixth edition of the CIOB Code of Practice for Project Management.

These six themes are interdependent with each other and are presented in Figure 1.3. While practitioners may take and explore any one theme on its own and in more detail than another, they must all be considered and integrated into the programme life cycle strategy.

Our aim here in this Code of Practice is not to provide a comprehensive review of these themes, but to provide an introduction to each of them and to point programme managers towards key industry and academic texts that can help practitioners develop their programme delivery strategies. Programme sponsors and programme managers must consider how and to what extent these themes are incorporated into programmes in the early inception, initiation and definition stages of the programme.

1.1.4 Developing capability

Achieving a client's strategic objectives via programme management is not just about delivering new client capabilities, but about how the programme organisation itself becomes capable of delivering these new client capabilities. This is because any form of programme organisation needs to 'become' capable and adapt those capabilities through the life of the programme if they are to turn inputs into outputs in order to deliver client objectives.

Capabilities do not 'just appear' because we have designed an organisation, written a set of management plans and procured necessary resources from different firms. Programmes, as a form of organising, need time and effort to reproduce programme-specific routines that are the building blocks of organisational capabilities. This is difficult when different firms and people move in and out of the programme and projects at different times in the life cycle. More importantly, the time needed for reproducing routines and building capabilities is often curtailed through price and time-driven procurement models.

The industry's focus on developing collaborative working practices is the way forward. These practices need to be developed in the early stages of the programme with a specific focus on the interface between the firms and the programme management team, identified as an area that lacks attention by the industry.[3] Through their procurement and delivery strategies, programme managers need to address the challenge of integrating together the hard transactional arrangements between the different organisations and developing the relational aspects of multiple participants working together, often for the first time and which change through the life of the programme. Programmes are often of a longer duration than projects, and this provides an opportunity for people-centred strategies of capability development to be put in place. This should focus on individual competency, skills and career development as well as building a collaborative team culture. As we say, these should not be independent of the commercial transactions between organisations and thus these issues need to be considered as a part of the procurement strategies developed for the programme.

1.1.5 Procuring value

Building on the foundational work of the Latham and Egan reports in the 1990s, there has been an increasing focus on collaboration and considerable effort put into understanding how we define, measure and procure value in its widest sense.[4,5,6] Such a shift in focus from the traditional price and time-driven approach to procurement rightly focuses programme management, in its early stages, on moving away from simply considering the outputs of the programme towards a more comprehensive view of the outcomes and benefits that the programme will deliver. This is not just for the client but for society more generally, and more specifically, the societies within which the programmes are being delivered.

---

[3] Addyman, S., and Smyth, H., 2023. *Construction Project Organising*. John Wiley & Sons, Ltd.
[4] https://www.gov.uk/government/publications/the-construction-playbook.
[5] https://constructioninnovationhub.org.uk/value-toolkit/.
[6] https://www.constructionleadershipcouncil.co.uk/news/procuring-for-value/.

The ability of the client organisation to understand their problem, define their requirements and develop the scope for any project or programme has been demonstrated as a necessary capability for the success of any project or programme.[7] A move to understand how clients procure the value to be derived from these front-end activities has developed in parallel with our understanding of the need for more collaborative approaches to the development and delivery of projects and programmes.

While traditional and design and build procurement models continue to dominate the industry,[8,9] what we are seeing is a need for client organisations to engage collaboratively with potential suppliers at the earliest stages of a project or programme so that the supply chain is able to become fully integrated in helping understand the need, scope and requirements.[10,11] This, in turn, drives the programme team to look at different ways in which they not just manage but also engage with and integrate programme stakeholders into this process.[12] This is also important for maximising the opportunity to create routines in the early stages of the programme, which in turn helps produce the necessary programme capabilities we talk about above. Furthermore, where programmes involve the application for statutory planning consent, integrating stakeholders and supply chains in the early stages helps mitigate future problems of being successfully granted that consent in a timely manner and for the planning conditions being integrated into supply chain contracts before full mobilisation for detailed design and construction commences. This is an area that is often neglected and delivered outside of the programme delivery organisation.

Such an approach is not without its challenges, as early commercial lock-in may lead to misalignment of behaviours. However, the rewards of taking such an approach, if managed carefully, have been shown to outweigh some of the negative aspects.

1.1.6 Digitalisation, visualisation and data

The digital revolution in society and construction more specifically continues at a pace.[13] In our first edition, we rightly drew attention to the rise of BIM as a digital approach that can help underpin collaborative working. In this second edition, we point programme managers towards a greater appreciation of how this digital revolution needs to be considered in two ways: digitisation (being the ongoing advancement of digital technologies in our industry) and digitalisation (being the capability of the programme organisation to become capable of adopting and applying these technologies).

What we have learned is that the rate of change in digital technologies and their application means that this is going to be dynamic and evolving throughout the life of a programme and is an important aspect of developing capabilities that we set above. Programme managers therefore need to not just build these technologies into their delivery strategies but also identify and implement the ways in which the

---

7   cf. NAO: Lessons from Major rail infrastructure programmes. https://www.nao.org.uk/reports/lessons-from-major-rail-infrastructure-programmes/.

8   Oyegoke, A.S., Dickinson, M., Khalfan, M.M., McDermott, P., and Rowlinson, S., 2009. Construction project procurement routes: an in-depth critique. *International Journal of Managing Projects in Business*, 2(3), pp. 338–354.

9   RIBA Construction Contracts and Law Report 2022. https://www.architecture.com/knowledge-and-resources/knowledge-landing-page/riba-construction-contracts-and-law-report-2022.

10   https://www.gov.uk/government/publications/the-construction-playbook.

11   Mosey, D., 2019. *Collaborative construction procurement and improved value*. Hoboken, NJ, USA: Wiley-Blackwell.

12   Di Maddaloni, F., and Sabini, L., 2022. Very important, yet very neglected: Where do local communities stand when examining social sustainability in major construction projects? *International Journal of Project Management*, 40(7), pp. 778–797.

13   https://www.ciob.org/industry/policy-research/resources/digital-construction?gclid=CjOKCQjwsIejBhDOARIsANYqkD1ijb4IWIFPbhyDOy_td7GUNBzXc3nt9NCFz36nuMzn1duVT8tagtYaAgVWEALw_wcB.

programme organisation will continue to evaluate and adapt its digital capabilities (both technology and personnel) during the whole life of the project. This requires some analysis and practical foresight for what might be at present and what might come in the future, as well as an appreciation of the different digital capabilities that exist in the wide range of supply chain partners that vary considerably between large organisations and small and medium enterprises (SMEs).

Furthermore, these digital technologies produce a lot of data, and programme managers need to be creative and innovative in the ways that organisational routines are reproduced in ways that enable teams to share this data and to 'visualise' and communicate on the performance of the organisation, both internally and externally.

1.1.7 Cultural identity and ethics

Programmes encompass a wide range of organisations and individuals that vary in their day-to-day practices and management styles. But all need to have ethical standards of trust and behaviour that are mutually acceptable, even if they contain variations, if their relationships within the programme are to be successful and sustainable. We know relationships are important, and relationships need managing. In this sense, programme managers need to reproduce practices and routines that facilitate ways for participants to be able to relate to each other in their professional roles.

For all the work that has been done on improving health and safety in construction and all the talk of being more collaborative, as an industry, we still have what has been termed a toxic culture[14] and lag behind other industries in terms of the wellbeing of the workforce.[15] Most notably, skilled construction and building trades rank first in the ONS (Office for National Statistics) statistics for in-work male suicide.[16] A situation that must change if we are to transform the industry.

The remedies for such problems are not easy to come by or implement, but as we have already pointed out in this Code of Practice, such problems cannot be approached solely from the top-down through the prescriptive applications of standards or rules at the institutional or firm level. Programme managers may use ideas that we have presented in the above themes for developing programme capabilities, collaborative procurement and a focus on social value as ways to approach such systemic industry problems. We see programmes and programme managers as a force for positive change in this area, as they arguably have the time and scope to positively influence the change that the industry needs in this area.

This will become ever more important as the digitalisation of the industry increases and we move towards more modern methods of construction. We will see changes in the way that workers from all professions and skill levels engage with the industry, and to ignore worker wellbeing in this move will perpetuate some of those negative practices that are seen to induce the current problems we have.

1.1.8 Sustainability and its governance

Environmental performance and impact, together with the other sustainability elements of 'economic' and 'social', are becoming increasingly important to both

---

14 Clegg, S., Loosemore, M., Walker, D., van Marrewijk, A., and Sankaran, S., 2023. Construction Cultures: Sources, Signs, and Solutions of Toxicity, Ch. 1, pp. 3–16. In: Addyman and Smyth (Eds), *Construction Project Organising*.

15 Xu, J., and Wu, Y., 2023. Organising Occupational Health, Safety, and Well-Being in Construction: Working to Rule or Working Towards Well-Being? Ch. 2, pp. 17–30. In: Addyman and Smyth (Eds), *Construction Project Organising*.

16 ONS (September 2021). *Suicide by Occupation, England: 2011 to 2021*. UK: Office for National Statistics: https://www.ons .gov.uk/peoplepopulationandcommunity/birthsdeathsandmarriages/deaths/adhocs/ 13674suicidebyoccupationenglandandwales2011to2020registrations.

clients and firms in the built environment. Programme aims need to consider both 'sustainability by the programme' (i.e. managing in a socially, economically, ethically and/or environmentally viable fashion) and 'sustainability of the project' (i.e. the ability of benefits to endure beyond the end of the programme/project).[17] This includes requirements on carbon emissions and energy consumption. In addition, it may also prescribe requirements for the environmental impact on local topographies or areas adjacent to related projects. It may determine outcomes in terms of associated communities, such as providing employment and training opportunities or the use of supply chains.

Increasingly, clients are setting out these requirements in contracts, which in turn pose a challenge for firms operating in the built environment due to the temporary and geographically dispersed nature of the projects that they deliver. What we have seen here is firms engaging with third-sector organisations (i.e. non-governmental, charitable) that can provide a bridge between firm strategies, client requirements and the local communities within which they work.[18] Clients and programme managers need to work together to develop and deliver requirements that can reasonably achieve these objectives within the bounds of individual programmes.

1.1.9 Platform thinking and modern methods of construction

Since the first edition of this Code of Practice, we have seen a growth in interest towards modern methods of construction (MMC)[19] and product platforms,[20] developed into what has been termed platform thinking. A recent study on platforms in construction[21] extended the understanding to four types of platforms: (i) platform organisations; (ii) product platforms; (iii) platform ecosystems; and (iv) market intermediary platforms. Due to the nature and scale of programmes in the built environment, platform thinking offers a new opportunity to bring about transformative change in the industry and offers new ways for thinking about how strategic objectives can be delivered both effectively and efficiently.

Central to this way of thinking is the digital revolution that continues to influence the industry, and while approaches to the application of platforms in construction are new, strategies for their implementation and application are growing. The transforming construction network + report presents two approaches to designing platforms:

(1) Top-down – conceived and designed from scratch;

(2) Bottom-up – existing structures analysed for common features.

As with any new approach, programme managers need to be cognisant of the existing capabilities of the organisations involved in a programme before seeking to push through a new initiative that may, if not duly considered, become more of a constraint than an enabler to better performance. We recommend the place to start here is the product platform rulebook developed by the Construction Innovation Hub.

17 Huemann, M., and Silvius, G., 2017. Projects to create the future: Managing projects meets sustainable development. *International Journal of Project Management*, 35(6), pp. 1066–1077.

18 Loosemore, M., Alkilani, S.Z., and Murphy, R., 2021. The institutional drivers of social procurement implementation in Australian construction projects. *International journal of project management*, 39(7), pp. 750–761.

19 Farmer, M., 2016. Modernise or die: Time to decide the industry's future. Construction Leadership Council. London, UK. https://www.constructionleadershipcouncil.co.uk/wp-content/uploads/2016/10/Farmer-Review.pdf.

20 The Construction Innovation Hub: The Value of Platforms in Construction. https://constructioninnovationhub.org.uk/media/rzwdinep/the-value-of-platforms-final-upload-april-2023.pdf.

21 Mosca, L., Jones, K., Davies, A., Whyte, J., and Glass, J., 2020. Platform Thinking for Construction, Transforming Construction Network Plus, Digest Series, No.2.

1.2 Applying programme management in practice

1.2.1 Introduction to programme management

Unlike projects and portfolios, programmes are created for the horizontal coordination of multiple projects, which may run sequentially or in parallel. From a business and customer perspective, a programme is designed to operate, learn and adapt in a dynamic environment of interrelated projects, people and organisations. Programmes do not necessarily have the strictly finite nature of a project.

An undertaking is considered and executed as a programme when some or all of the following are considered relevant:

- the delivery criteria may or may not be fully known, defined or approved;

- the undertaking requires a high level of regulated governance;

- risk and opportunity across related projects are best managed through a central function;

- achievement of the overall outcome required necessitates a number of related projects, each demanding different specialist skills, expertise or organisational approaches;

- the size, complexity and uncertainty of the undertaking are such that delivery is best approached by creating a number of projects;

- the delivery skills required are beyond the organisational and contractual arrangements for one team;

- the geographic spread of the undertaking makes it uneconomic or infeasible to have one project;

- time or cost constraints mean that it is uneconomic or infeasible to have one project;

- there is a requirement to manage interdependencies between projects;

- there is a requirement to manage conflicting priorities and resources across projects.

In the context of construction, CIOB defines a programme in the following way (see Figure 1.4):

> A programme is a collective of related projects coordinated to achieve desired benefits more effectively than managing them as a group of individual projects.

A programme therefore comprises a collective of related projects that are limited in time and designed to individually deliver agreed-upon objectives and that produce and deliver a product, service or outcome. Projects may be internally delivered or outsourced to specialist suppliers and/or contractors, or a mix of both. The coordinated manner in which they are managed delivers programme benefits that are greater than the sum of individual project benefits if they were not coordinated at the programme level.

1.2.2 Types of programmes

The task of programme management is to create and coordinate this collective of related projects in order to deliver programme benefits, which would not be as

Programme

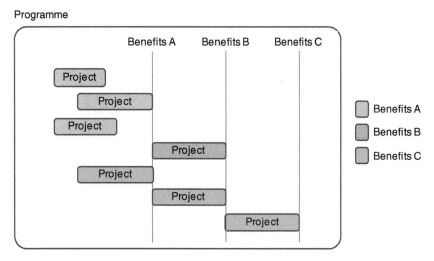

Figure 1.4 Programme, projects and benefits.

achievable if they were managed as a group of individual projects. Success of a programme is thus dependent on a programme team's ability to deliver those benefits. Programme management brings reality and method to strategic planning and to the delivery of strategic change with the right strategic objectives phased and delivered on cost, on time, on quality and on benefits.

In the context of the built environment, programmes may be categorised into three key types, depending on the driver for change. These are:

(a) Type 1: **Vision-led programmes** that are necessarily top-driven and set out to meet a particular strategic vision or need;

(b) Type 2: **Emergent programmes**, where the organisation recognises that a group of existing projects would be better managed together (a bottom-up approach); and

(c) Type 3: **Compliance programmes** that are created in response to internal or external stimuli, often generated outside the control of the organisation.

All programmes, regardless of the driver and type, introduce change: internal or external or even both internal and external.

Example 1: Vision-Led Programmes

Reducing the carbon footprint of existing and new hospitals is a vision that requires a number of diverse retrofit and new-built projects according to commonly accepted standards, staff training, logistics and other projects. Mobilisation of vast resources across geographical regions calls for coordination far beyond the needs of a single project.

Example 2: Emergent Programmes

An organisation facing similar problems in several existing projects recognises that they need to be addressed across all projects. For example, some common factors that lead to delays and other types of losses need to be addressed holistically, and solutions developed and implemented in all projects. A programme could then be formed by pulling together existing projects to develop and implement solutions across all projects in the programme.

Example 3: Compliance Programmes

A change in facilities-related legislation could force an organisation with a large number of facilities to form a programme to implement the necessary changes across all facilities. Large organisations where compliance-related changes normally lead to the formation of programmes designed to implement the necessary changes normally need to maintain in-house programme management capacity and capability.

1.2.3 Programme management process and stages

Programmes can be of different sizes and complexity. Some will include a large number of projects over many years with major milestones. Other programmes will be fairly small, with only a few projects. Regardless of their size and complexity, programmes in the built environment are best served by following the traditional linear delivery life cycle model for projects, which offers a robust model for programme planning and delivery.

For the application of this model to be effective, programme teams will need to adapt their organisation and management approach to anticipate and integrate changes within their financial, contractual and physical constraints, as the programme moves through its life cycle. More specifically, programme teams will need to plan for and manage the transitions from one life cycle to the next, setting out changes to organisation and governance arrangements.

Within a programme, different project management approaches, such as waterfall or agile, may be applied to individual projects where appropriate. Innovation and creativity through clarity of strategic intent and creative design coupled with adapted controls (risk, opportunity and change management) are critical to successful scope delivery (on time, to budget and quality) and benefit realisation. The programme and organisational structure (including resource levels) will be gate-controlled and evolve through each stage of the programme.

The following stages determine a framework for through-life management of programmes:

- Stage A: Programme inception
- Stage B: Programme initiation
- Stage C: Programme definition
- Stage D: Programme implementation
- Stage E: Programme benefits realisation and transition
- Stage F: Programme closure

The purpose of each stage, the key activities of each stage, and the key roles and responsibilities of each stage are set out in detail in Chapters 5–10 and represented here graphically in Figure 1.5.

Appendix A provides an overview of the key artefacts that are produced for each of the stages and how the entire process relates to the RIBA[22] plan of work 2020.

1.3 Programme organisation

As described above, programmes may be vision-led, emergent or compliance-based (driven by external pressure). That means some programmes will be internal, others will be developed on behalf of an external client, and some will be a mix of

---

[22] https://www.architecture.com/knowledge-and-resources/resources-landing-page/riba-plan-of-work.

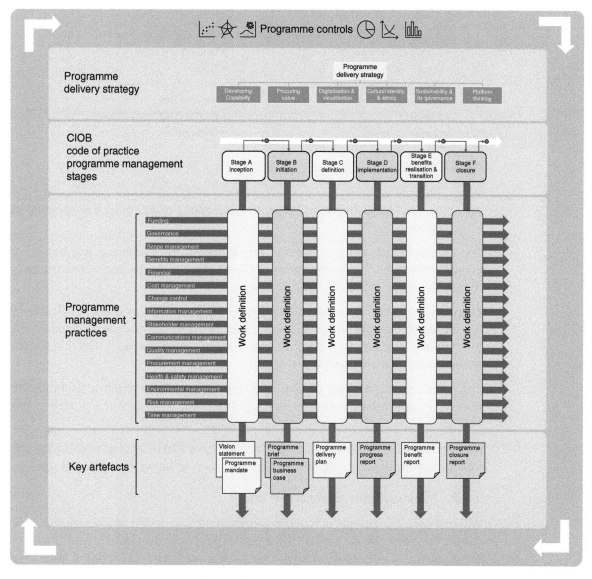

Figure 1.5 CIOB Programme management structure.

both. Regardless of the case, a programme is always developed to deliver benefits to a client (internal or external). Although the nature of their businesses and their approaches and structures in the private and public sectors are diverse, it is common that interim or final outcomes and deliverables will need to be reported to a board of directors.

1.3.1 Types of clients who may initiate programmes

In this Code of Practice, a client is an organisation that procures the services of professionals (either internally or externally) to initiate, develop and deliver a programme of projects. In the built environment, client organisations broadly range from:

(1) clients that undertake construction as a one-off event to extend or enhance their business activity;

(2) those that undertake programmes and projects as their core business activity;

(3) those that do programmes and projects alongside their core business activity; and

Typical public sector clients
• Central Government
• Local Government
• Government departments
• Health
• Transport
• Energy

Typical private sector clients
• Banking and finance
• Retail
• Hospitality and leisure
• Manufacturing
• Food and beverage outlets
• Developers

Figure 1.6 Typical organisations.

(4) those that are established for the single purpose of delivering a programme of projects.

The organisational and management arrangements for developing and delivering projects vary hugely across this array of clients, from large, dedicated departments to a single appointed responsible person.

Client organisations undertaking programme management in the built environment typically come from the private and public sectors, although they may also come from the third sector (i.e. charitable or nongovernmental organisations). Figure 1.6 lists typical public or private sector clients.

The client organisation will usually 'own' the programme and the projects and is likely to have obtained their own funding. The client organisation will appoint the programme sponsor as the person who represents the client.

All programmes differ in structure, but they will have similar characteristics. The following sections describe the attributes of the key programme management roles.

1.3.2 Key programme management roles

Whether controlling single or multiple programmes, clients will allocate a number of roles overseeing the delivery of the programme. Depending on the type of client, these roles may be sourced internally from the client organisation, or the client may procure this expertise from an external organisation. The two primary roles that must be appointed are programme sponsor and programme manager, both supported by management boards and a programme management office, as follows:

1.3.3 Programme sponsor

A programme sponsor outlines programme vision, objectives and benefits. Directly responsible for developing the vision statement into the programme mandate, a programme sponsor's key responsibilities include:

• strategic direction and fit with the overall business strategy;

• releasing the required resources;

• ensuring programme stability in terms of time, budget and scope;

• championing the programme at the most senior level;

• providing high-level feedback on programme progress.

The role of programme sponsor is a very senior position requiring visionary capabilities and competent leadership skills acquired from leading diverse senior teams.

In some organisations the programme sponsor role may be known as the senior responsible owner or the programme director.

1.3.4 Programme sponsor's board

The programme sponsor's board will have the authority to make key decisions and commit expenditures on the programme on behalf of the client organisation. It will consist of executive-level individuals who are heads of functions of either the sponsoring organisation or of the final commissioned enterprise. Membership of the programme sponsor's board needs to be ratified at the highest level, by the executive board or chief executive officer of the sponsoring organisation.

Throughout the programme, the role of the programme sponsor's board is to provide the overall strategic direction, support the programme sponsor in the implementation of the programme, ensure that adequate resources are available to the programme, monitor the programme's progress towards achieving the required outputs, outcomes and benefits, facilitate the resolution of any major issues, determine when the programme's objectives have been achieved and ratify closure of the programme.

1.3.5 Programme manager

The programme manager coherently manages programme stages, reports to the programme sponsor and is responsible for the delivery of the proposed change. Key responsibilities include:

- developing and maintaining a project-supportive programme environment;
- working with the programme sponsor to ensure the programme is delivered on time, within budget and scope;
- managing the programme management office;
- delivering programmes successfully in terms of agreed objectives and identified benefits (i.e. programme finances and benefits).

The programme manager is a senior appointment. The person in that role must have the necessary skills to implement the programme. In addition to understanding project management skills, a programme manager should have:

- sound understanding of business case development;
- good knowledge of key programme-level financial and business indicators;
- senior-level credibility to effectively support project teams;
- excellent stakeholder management skills.

1.3.6 Programme management board

The programme management board is composed of senior managers of the programme management structure and provides advice and support to the programme manager. Its key responsibilities include:

- reviewing progress;
- highlighting and resolving any issues that may be hindering progress.

1.3.7 Programme management office

The programme management office is a central support unit that supports the programme manager in overseeing day-to-day operation of a programme. It includes a senior manager and other specialist staff to carry out the functions required to:

- develop the programme delivery plan;
- develop and maintain standards;

- establish the governance controls;

- manage programme documentation;

- enhance capability to deliver the programme.

1.3.8 Wider programme management team

The programme sponsor and programme manager will be supported by a wider programme team. The roles, responsibilities and design of the programme team will vary from client to client and depend on the type of programme. In this Code of Practice, we put forward the following key roles:

- Programme business change manager;

- Programme benefits realisation manager;

- Programme financial manager;

- Head of programme management office;

- Programme risk manager;

- Programme scheduling manager;

- Programme cost manager.

In each programme stage and Appendix D, we set out these roles and responsibilities and their key competencies.

1.3.9 Stakeholders

Stakeholders include persons and organisations that have an interest in the strategy of the organisation and programme, have an impact or are impacted by a programme.

Programmes and organisations have to identify all stakeholders and assess the level of power they hold to affect the decisions and outcomes of the programme.

Stakeholders can be divided into two groups:

- Internal stakeholders: members of the organisations and those with an economic or contractual relationship with the programme;

- External stakeholders: those with interest in the organisation and programme activities or those impacted by the activities in some way, such as governments, the public, interest and pressure groups, media and news organisations, local communities and statutory authorities.

A list of common stakeholders may include the following:

- general public (people who are only indirectly affected by a programme but who may have a significant influence on its realisation);

- community (people who are directly affected by a programme through their geographic proximity to programme works);

- client employees (delivered changes and benefits will directly impact employees of the client organisation);

- shareholders (individuals or legal entities owning shares in the client organisation undergoing change that are affected by the business change);

- end users (those who will ultimately work in new facilities provided or who will be the beneficiaries of the outcomes of the programme);

- customers (customers of client organisations who will be affected by the business change);

- statutory and regulatory authorities (most programmes will be subject to a range of organisations that will impose restrictions on the way they can be implemented);

- interest groups (the members of which share common interests and control some area of activity, e.g. nonprofit organisations and voluntary organisations).

Programmes need to identify and map the stakeholder landscape to engage and communicate effectively. To assess the level of engagement required and the impact stakeholders will have in meeting the objectives of the programme, a stakeholder map should be developed. This map will define the different tiers of stakeholders according to their potential to affect the reputation or the progress of the project or organisation, with the programme at the centre.

1.3.10 Portfolio management

In the eventuality that there are a number of projects and programmes running in parallel within an organisation, the organisation will often utilise a portfolio approach to govern and administer the initiatives, projects and programmes to identify and manage priorities.

The portfolio management approach will aim to understand the current strategic intent of the organisation and will determine the optimum spectrum of programmes and projects that would provide the most effective and efficient way of achieving the strategic vision by balancing the resources, risks and benefits sought.

Typically, in any organisation, portfolio management is an ongoing activity, for unlike most programmes and all projects, it will not have a defined end date.

The administration, management and governance of portfolios follow principles similar to those of a programme; however, unlike programmes, the procedures are open-ended and subject to continual reviews at the highest level of the organisation.

Large organisations may have multiple portfolios, in which case an additional layer of management and governance will be necessary between the senior decision-makers and the portfolio management levels for reporting and administration purposes.

2 Stage A: Inception

2.1 Purpose of this stage

The purpose of Stage A is to determine whether the strategic aspirations of an organisation can be achieved by executing a programme of work. This stage must include due consideration of the six core themes that must be integrated into the management of a programme, as set out in Section 1.1.3 in Chapter 1.

Please refer to the programme delivery matrix (PDM) in Appendix A to understand the key roles and artefacts that comprise the CIOB programme life cycle.

2.2 Key activities of this stage

The vision statement is the first document that is produced by the organisation interested in implementing a programme. The vision statement describes the original intention, benefits and outcomes that are required by the programme.

Once approved, the vision statement becomes the basis for the development of the programme mandate (see Figure 2.1).

2.2.1 Vision statement

The process of determining the need for a programme is likely to be complex and lengthy. This is because programmes by their nature are expected to be large enterprises, involving a wide mix of interested parties, large expenditure of capital and resources and creating a large environmental and societal impact.

The inception stage will be undertaken by the client organisation, which itself may comprise a number of separate legal entities.

At a senior executive level in the client organisation, there will be consideration of the need for a business change, or for the creation of a new enterprise or capability, that arises out of the strategic business objectives of the organisation.

This corporate aspiration is described by a vision statement, which is produced at an executive level and sets out the intent and benefits being sought. The vision statement is likely to be subject to board-level approval and authorisation.

The key attributes of a vision statement can be summarised as:

- focused: simple, clear and concise
- motivational and inspirational
- feasible, with realistic and achievable goals

Code of Practice for Programme Management in the Built Environment, Second Edition. The Chartered Institute of Building.
© 2024 John Wiley & Sons Ltd. Published 2024 by John Wiley & Sons Ltd.

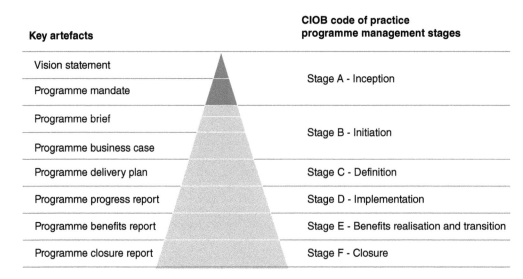

Key artefacts		CIOB code of practice programme management stages
Vision statement		Stage A - Inception
Programme mandate		
Programme brief		Stage B - Initiation
Programme business case		
Programme delivery plan		Stage C - Definition
Programme progress report		Stage D - Implementation
Programme benefits report		Stage E - Benefits realisation and transition
Programme closure report		Stage F - Closure

Figure 2.1 Key artefacts: Stage A (Inception).

- unambiguous and collectively understood by all in the same way

- free from technical jargon and uses common and plain language

2.2.2 Programme mandate

When the programme sponsor is appointed, one of the first actions is to expand on the vision statement by setting out in greater detail what the programme needs to achieve in terms of outcomes and what the programme seeks to deliver. This will be the focus of the programme mandate.

Depending on the nature of the programme, defining it may be obvious and straightforward or it may be highly complex, requiring development in association with parts of the client organisation and possibly with external parties.

2.2.3 Managing strategic change

Introduction

Programme management is a systemised and structured approach designed to realise strategic objectives and manage the risks of delivering them.

Programme management brings reality and method to the aspirations of strategic planning and to the delivery of change with the right objectives phased and delivered 'on cost, on time, on quality and with benefits'.

A programme is a temporary organisation designed to operate, learn and adapt in a complex environment of interrelated projects, people and organisations.

Programme delivery in context of the built environment is a circular process running from strategy to benefits realisation, framed between:

(i) business as usual – strategising and realising benefits while operating a business/an organisation – and

(ii) programme environment – planning and managing a programme to deliver outcomes and benefits (see Figure 2.2).

Strategic planning

Strategic planning is typically carried out at the organisational governance level. It is at this level that the strategic direction for the organisation is set. Normally initiated

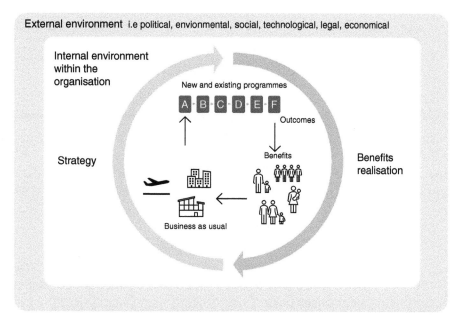

External environment i.e political, envionmental, social, technological, legal, economical

Internal environment within the organisation

New and existing programmes

A B C D E F

Outcomes

Benefits

Strategy

Business as usual

Benefits realisation

Figure 2.2 Programmes and their environments.

at a senior level, it is a process that defines the business intent, formulates the vision statement and captures the strategic changes to be delivered that will have positive economic, environmental and social impacts.

Strategic changes are defined and selected through the strategic planning process and the analysis of macro-environmental scenarios in order to produce qualitative and quantitative benefits, adding value to the expected return of this investment. For each environmental scenario driving a strategic change, the following questions will be considered:

- What political, environmental, social and technological trends are you noticing?
- What are the related issues or challenges?
- What advantages or opportunities are there?
- What impact might these have on the organisation?

A number of examples are listed below for each type of change:

- To enter new geographical market
- To renew equipment fleet
- To become leader in the use of artificial intelligence technology
- To develop strategic capability in programme management
- To implement operational waste regulation
- To implement carbon emissions regulation
- To adopt new licensing requirements
- To increase passenger capacity
- To deliver a major sporting event
- To improve flood control
- To deliver alternative energy generation sources
- To reduce patient waiting time for hospital procedures

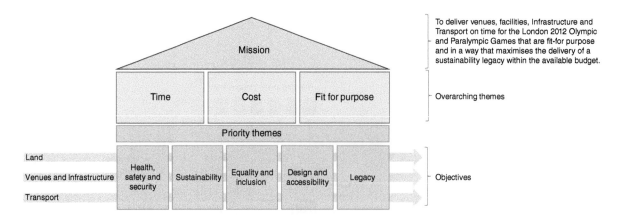

Figure 2.3 Overall strategic approach, UK Olympics.

An overview of the overall strategic approach (see Figure 2.3) adopted by the Olympic Delivery Authority to deliver the 2012 London Olympic Games' physical assets is summarised below.

Each project – land, venues and infrastructure, transport – was aligned to overarching and priority themes and a set of objectives established for programme delivery and decision-making purposes against specific reporting metrics.

The Olympic legacy programme was designed to be delivered through three separate organisations funded in majority by the UK government as follows:

- Olympic Delivery Authority (ODA) – to design and build the venues for games and legacy purposes
- London Organising Committee of the Olympic and Paralympic Games (LOCOG) – to run the Olympic Games
- Olympic Park Legacy Company (OPLC) – to transition the assets into legacy mode

Business objectives

To deliver a successful programme, managers need to understand and consistently communicate their organisation's values and vision for strategic change. They also need to translate these into business objectives that are SMART (Specific, Measurable, Achievable, Realistic and Time bound) in order to deliver sustainable benefits. The following is an example.

> A building manager has a strategic objective to generate revenue from a number of vacant sites. At the operational level, construction projects may deliver commercial and residential buildings to the building owner; these can form a programme of works containing multiple projects. Once the projects are completed, the owner has the capability to generate profit, which would enable achievement of the programme benefits. Only when the buildings are leased or sold can the benefit be realised and measured. When the expected revenue (as identified during the strategic decision-making stage) has been achieved, the strategic objective is deemed to have been met, thus effecting the completion of the programme.

An effective way to capture strategic objectives and filter down this information in a programme and an organisation's functional areas is illustrated in Figure 2.4.

Type of change	Strategic change	Strategic objectives
Internal changes	To enter new geographical market	Enter x new market and generate £x million in revenues by year
	To renew equipment fleet	Review supply chain and renew xx% by year xxxx
	To become leader in use of a new technology	Use BIM for all projects over £xxm by year xxxx
	To develop strategic capability in programme management	Select and up-skill xxxx senior managers/year
External changes	To implement operational waste regulation	Recycle 100% of project waste by year xxxx
	To implement carbon emission regulation	Reduce carbon emission by 20%year on year
	To adopt new licensing requirements	Implement new licenses requirements on all sites within x months
End user/client lead (i.e. client lead strategic change requiring new physical assets)	To increase passenger capacity	Build new airport, train line, port, etc. by year xxxx
	To deliver a major sporting event	Deliver on time and on budget within scope and safely
	To improve flood controls	Reduce flood risk in x area by xx%
	To delive ralternative energy generation sources	Generate xx% of country energy needs from alternatives sources by year xxxx
	To reduce patient waiting time for hospital admission	Reduce waiting time to xx weeks for specific clinical cases by year xxxx

Figure 2.4 Strategic change and strategic objectives by change type.

Business change process

For client-led programmes (i.e. delivering multiple physical assets or complex long-term contracts), the focus of the following sections, the traditional linear delivery cycle for major projects offers a robust model for delivery. Using this model, an effective programme will anticipate and integrate organisational changes within financial, contractual and physical constraints as time passes. Innovation through clarity of strategic intent and creative design coupled with adapted controls is critical to successful scope delivery and benefits realisation. The programme and organisational structure will be gate-controlled and will evolve through each delivery phase (see Figure 2.5).

Funding policy

Sources of funding for a programme may come from single or multiple (internal or external) organisations or business units contributing to the programme budget and anticipating a benefit from the strategic change. Not all the projects within the programme may have the same funders, given the potential time variances. Hence, the number and range of funders may change over time; typically, these may include banks, pension funds and insurance companies, together with investors from in-country or other international sources. This is likely to depend on where the individual projects are located. The nature of the projects themselves may attract different types of investors and funding organisations.

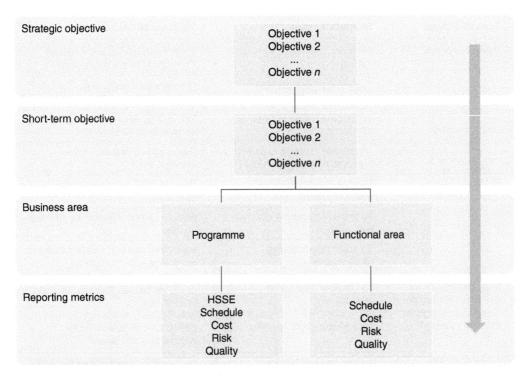

Figure 2.5 Strategic objectives alignment.

The funding policy for the programme typically depends on the nature of the programme and whether it is in public sector or private sector. At the inception stage, the key decision for the type and nature of funding may not be fully detailed; perhaps, neither would the budgetary requirements or cash flow. Depending on the outputs and outcomes, it is possible that different streams of funding may have to be pulled together for different projects and outcomes.

Some of the initial considerations in relation to funding arrangements may include the following:

- Is the proposed programme a strategic fit for the organisation?

- Does the work require long-term and extensive support?

- Are there adequate considerations for lowering the financial risk, thereby reducing the cost of money?

- Is the proposed programme in an area of high strategic priority for a large investment?

- Is there a case for a major investment in the context of the overall portfolio and budget?

- Does the programme mandate generate confidence that the programme has the potential to successfully manage and deliver the benefits?

- What is the quality of the proposed programme?

- Are the risks clearly defined, with mitigation measures proposed?

- Are there contingency options in place?

- Are the delegated authorities clearly defined?

- How significant is the proposed programme in terms of its potential impact?

- Are there clear and deliverable benefits?

- Is it important to pursue this programme now?

- Is the vision and mandate realistic in their time frames and proposed resources?

- How convincing and coherent is the overall proposed approach?

- Has the organisation demonstrated a clear commitment to the proposed programme and the desired benefits?

- Are there any internal or external dependencies or funding?

- Does the programme represent good value for money?

- Are the proposed arrangements for stakeholder understanding of this programme appropriate and sufficient?

Not all of the above considerations may be fully determined at this stage; however, these are some of the key criteria that the programme brief (see Stage B) must be able to satisfy in order to secure funding.

2.3 Key roles and responsibilities of this stage

The inception stage requires the appointment of the programme sponsor by the client organisation; the programme sponsor has total overall responsibility for the programme and the achievement of the required outcomes. Adding detail and clarity to the initial aspiration of the client organisation is carried out principally by the programme sponsor.

During the course of this stage, the programme sponsor will acquire sufficient knowledge of the requirements of the proposed programme to determine the appropriate membership of the programme sponsor's board. This board will provide strategic direction to the programme and will be the body ratifying key approvals and decisions on behalf of the programme business partners (see Figure 2.6).

2.3.1 Client organisation

The client body that has decided to embark upon a major business change selects and appoints a suitably experienced individual who has the business, technical and managerial knowledge and skills to direct the successful delivery of the anticipated outcomes. The appointment of the programme sponsor should be accompanied by the client's production of a vision statement setting out what the programme needs to achieve.

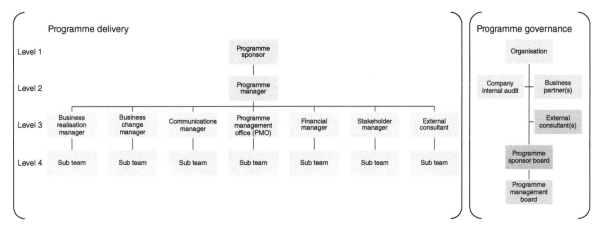

Figure 2.6 Typical programme organisation.

2.3.2 Business partners and funders

Those organisations that have a business or financial interest in the outcome of the programme assist the client in developing the vision statement and in providing advice to the programme sponsor in developing the programme mandate.

2.3.3 Programme sponsor

For very large programmes, it may be necessary for the role of the programme sponsor to be provided by a group or even by an organisation. In this situation, there still must be a lead role that carries the ultimate responsibility.

Key activities addressed by the programme sponsor to produce the programme mandate include the following:

- Reviewing the vision statement with the client
- Reviewing the vision statement with the programme business partners
- Breaking down the overall outcome into achievable objectives
- Engaging with external parties/key stakeholders who have specialist knowledge that helps to inform the programme mandate
- Determining the way the programme will be managed
- Preparing the programme mandate document

In relation to the programme sponsor's board, the programme sponsor needs to:

- Make recommendations to the client organisation on the proposed composition of the programme sponsor's board
- Develop terms of reference for the function and operation of the programme sponsor's board
- Oversee the formation of the programme sponsor's board
- Hold an induction meeting with the newly formed programme sponsor's board

Following the production of the programme mandate, the programme sponsor has further activities to undertake:

- Presenting the programme mandate to the programme sponsor's board
- Securing the approval of the programme sponsor's board for the programme mandate
- Developing terms of reference and plan for carrying out Stage B: Initiation
- Securing the approval of the programme sponsor's board to proceed to Stage B (including approval to commit the resources to achieve Stage B)

2.3.4 Programme sponsor's board

Once formed, the programme sponsor's board will be asked by the programme sponsor to review and ratify the programme mandate, together with the terms of references and time schedule indicating how the next stage will proceed. The programme sponsor's board approval allows the programme sponsor to proceed to the next stage of the programme. During inception, the programme sponsor's board undertakes a number of activities, including:

- Ratifying the selection and appointment of the business change manager
- Providing advice and input into the programme mandate

- Resolving issues raised by the programme mandate

- Ensuring the proposals contained in the programme mandate are consistent with the requirements of the functions/organisation that they are representing

- Reviewing and giving approval to the programme mandate

- Giving approval to proceed to Stage B

3 Stage B: Initiation

3.1 Purpose of this stage

The purpose of Stage B is to develop detailed proposals for the programme business case to determine what the programme will be able to deliver and make an informed judgement regarding its financial viability. This stage must include how the programme will deal with the core themes that must be integrated into the management of a programme.

Please refer to the programme delivery matrix (PDM) in Appendix A to understand the key roles and artefacts that comprise the CIOB programme life cycle.

3.2 Key activities of this stage

The required programme outcomes established during inception are developed through a more detailed analysis leading to the development of the programme brief. A programme business case is then compiled from the programme brief to determine the financial viability of the programme (see Figure 3.1).

3.2.1 Programme brief

The programme brief is developed from the programme mandate and vision statement, and covers the following aspects:

- Statement of intended outcomes
- Statement of the benefits required from the programme
- Awareness of how these benefits will be measured
- Consideration of the strategy for delivering the programme
- Indication of the organisational structure necessary to deliver the programme
- Statement of all costs of delivering the programme
- Statement regarding the funding mechanism
- Outline of time schedule for the achievement of the programme, including any critical dates, interdependencies between projects and any external dependencies or dates
- Identification of any external processes, procedures or approvals that will be relevant to the programme
- Statement of any key risks, issues, assumptions and constraints that have the potential to impact the delivery or outcome of the programme and its projects

Code of Practice for Programme Management in the Built Environment, Second Edition. The Chartered Institute of Building.
© 2024 John Wiley & Sons Ltd. Published 2024 by John Wiley & Sons Ltd.

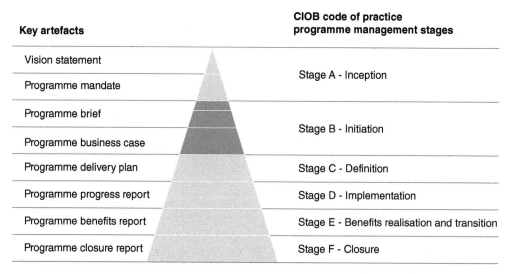

Key artefacts	CIOB code of practice programme management stages
Vision statement	Stage A - Inception
Programme mandate	
Programme brief	Stage B - Initiation
Programme business case	
Programme delivery plan	Stage C - Definition
Programme progress report	Stage D - Implementation
Programme benefits report	Stage E - Benefits realisation and transition
Programme closure report	Stage F - Closure

Figure 3.1 Key artefacts: Stage B (Initiation).

Who will assist with the preparation of the programme brief, other than the programme sponsor, will vary from programme to programme and will depend on factors such as the sensitivity (commercial, corporate or political) of the programme, the complexity of the programme and the degree of knowledge and expertise available within the sponsoring organisation.

In some circumstances, it may be possible for it to be produced by the programme sponsor, while in others, it may require the assistance of external advisers or the early involvement of some of the key members of the programme management team, such as the programme manager, programme financial manager or head of the programme management office.

At some point during the initiation stage, it will be necessary for the programme sponsor to have the specialist knowledge provided by somebody who has a thorough appreciation of what the final outcome of the programme needs to be. A business change manager views the development of the programme from the perspective of the final end-state. The business change manager will be appointed from within the client organisation.

When it is completed, the programme brief will be approved by the programme sponsor's board.

3.2.2 Programme business case

Based on the information contained in the programme brief, an investment appraisal is carried out balancing the expected benefits with the potential risks, opportunities and threats of delivering the programme. This information represents the programme business case. This is a document that will be used throughout the programme as a mechanism to verify that the deliverables being achieved are aligned with the programme outcomes.

Preparation of the business case is the responsibility of the programme sponsor, but it is expected that the sponsor will require assistance from financial and investment specialists.

When the programme sponsor considers that the business case presents a viable programme, it is submitted to the programme sponsor's board for their review and approval. By signing off on the business case, the programme sponsor's board is confirming they are satisfied the programme can proceed to the definition stage (Stage C).

As this approval commits a significant level of resources and expenditure in some instances, it may therefore be necessary to refer the business case to the client organisation's executive board in order to obtain the instruction to proceed to the next stage.

The programme sponsor's board should also be asked to ratify the terms of reference and time schedule indicating how the next stage will be delivered.

3.2.3 Benefits management

A generic approach to programme benefit management consists of three phases:

1. Capabilities phase: building and delivering the programme

2. Transition phase: transferring and operating the asset

3. Benefits phase: realising the benefits

Benefits delivery means achieving the desired outcomes identified in the business case. The key steps are benefits identification, benefits management and benefits realisation. To deliver benefits successfully, benefits need to be measurable outcomes and fall into one of the five categories, as identified in Figure 3.2.

Stage 1 – Benefit identification methods

The theme for improvement is a top-down method of articulating the focus areas in which the organisation wants to make a change. They give a clear sense of the areas of change and a broad understanding of where benefits will be seen (see Figure 3.3).

A more detailed approach to identify benefits and their measures is through a tree structure. A tree starts from the driver for change, and through a deductive, analytical and structured approach, processes through a series of options of potential benefits by category. Sometimes Ishikawa diagrams can be used to identify benefits.

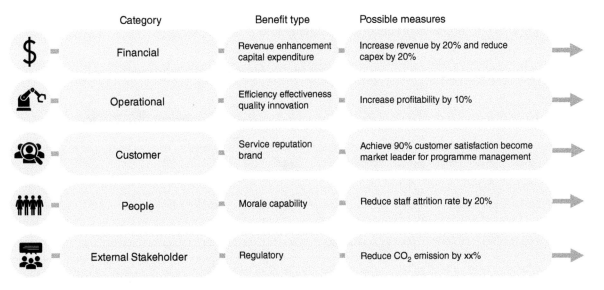

Figure 3.2 Benefit categories.

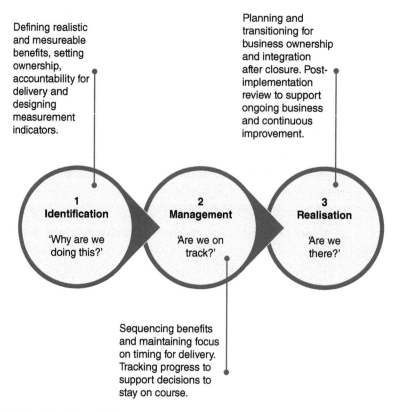

Defining realistic and mesureable benefits, setting ownership, accountability for delivery and designing measurement indicators.

Planning and transitioning for business ownership and integration after closure. Post-implementation review to support ongoing business and continuous improvement.

1 Identification

'Why are we doing this?'

2 Management

'Are we on track?'

3 Realisation

'Are we there?'

Sequencing benefits and maintaining focus on timing for delivery. Tracking progress to support decisions to stay on course.

Figure 3.3 Benefit delivery in three stages.

Stages 2 and 3 – Benefits management and realisation

Once all benefits are captured and defined, they can be held in a central database and summarised to include definitions by category, delivery measure and timescale for delivery and owner. For benefit tracking, a graphical chart can be used to represent and track benefit realisation over time (see Figure 3.4).

3.2.4 Feasibility study

The feasibility study should thoroughly examine all the issues and include the following:

- Appraisal of business alternatives with respect to meeting the identified need
- Appraisal of opportunities generated

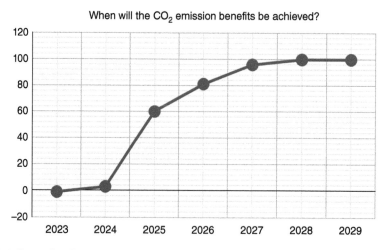

When will the CO_2 emission benefits be achieved?

Figure 3.4 Example of representation of benefits realisation over time.

- Assessment of high-level risks and potential mitigating factors

- Probability of successful delivery, at both programme and project level

- Funding options and scenarios

- Benefits appraisal

- Resources appraisal

- Any fundamental assumptions made and appraisal of the assumptions

- Demonstration of due diligence

It is important to note that once the programme brief is set, in particular for large and complex programmes, a scoping or options appraisal or even a pre-feasibility study may be necessary to narrow down the possible options and scenarios prior to commissioning a detailed feasibility study.

The feasibility study report should consider the programme brief thoroughly and specifically comment on the following:

- Technical viability

- Financial viability

- Benefits viability

- Contingency options

- Fitness for purpose

The feasibility study report, along with its recommendations, will constitute the backbone of the business case.

3.2.5 Funding arrangements

The initial investigations regarding the nature of funding will essentially determine the potential sources, availability and governance. An initial scoping study must be undertaken to determine whether the availability of funding is in keeping with the forecasted cash flow for the programme.

For private sector projects, commitments must be sought from the fund holders (and the shareholders or directors of businesses) to ensure that there is sufficient funding for the life of the programme. It is often the case that initial funding is released to kick-start the programme, and any further subsequent funding is conditional to trigger key events or achievement of early benefits (typically linked with revenue generation).

In certain instances, it is also advisable to ascertain the provenance of the funding to ensure that it is complying with regulatory and legal requirements.

Large public sector programmes often work with the private sector to create procurement alternatives, i.e. public–private partnership (PPP) or project finance schemes, as sources of programme funding. In this scenario, the key steps are the following:

- Implementation of the PPP/project finance/commercial policy

- Managing PPP and project finance projects within programmes

- Controlling the quality of PPP and project finance projects through procurement

- Supporting the transition through operational PPP projects to ensure that they achieve their benefits

- Managing the market of the operators and investors

- Embedding continuous improvement in PPP projects and programmes

In order to deliver the above functions, organisations are encouraged to ensure that the following enablers are considered at the initiation stage:

- A clear mandate of roles and responsibilities: There should be clarity regarding who is doing what, what the delegated limits are and who is authorised to make decisions

- Adequacy of resources – capacity and competency: The delivery team has the resources collectively experienced in PPP, commercial, financial, technical and sector-specific programmes and projects

- Strong relationship – intra- and inter-organisational: Strong relationships are needed within the delivery team and within the key organisations involved

A further option available to public sector organisations is self-financing, where funding could be available through borrowing (provided such borrowing is affordable, prudent and sustainable over the medium term), capital receipts (by selling fixed assets), capital grants (from various central and local government funding), revenue contributions or capital reserves (with funding earmarked for specific capital programmes).

3.3 Key roles and responsibilities of this stage

The key participants in this stage are the programme sponsor and business change manager, who work together to develop the details of the proposed programme such that its viability, in the form of a valid business case, can be demonstrated to the programme sponsor's board.

It is likely that during this process, the programme sponsor will require the assistance of the programme manager to ensure that the assumptions made regarding programme implementation are appropriate and achievable.

3.3.1 Programme sponsor's board

This programme sponsor's board will conduct the following in order to develop the programme brief:

- Selecting and appointing the programme business change manager

- Selecting and appointing the programme manager

- Ensuring, with the business change manager, that the benefits identified as being delivered by the programme are compatible with the business requirements of the client

- Providing initial consideration of the governance policies for the programme

- Developing the outline strategy for implementing the programme

- Providing initial consideration of likely timescale

- Providing initial consideration of likely cost

- Providing initial consideration of major constraints and risks

- Reviewing the client's funding options

- Producing the programme brief

- Presenting the programme brief to the programme sponsor's board

- Securing programme sponsor's board approval of the programme brief

Approval of the programme brief allows the programme sponsor's board to proceed with a series of activities related to developing the programme business case:

- Review, in conjunction with the business change manager, the identification of the benefits to be delivered by the programme

- Review, in conjunction with the programme manager, the methodology for delivering the programme

- Review with a specialist adviser the available funding options

- Develop a funding strategy

- Review any relevant lessons learned from previous projects or programmes

- Oversee the production of the programme business case

- Present the programme business case to the programme sponsor's board

- Secure approval of the programme sponsor's board for the programme business case

- Prior to completion of this stage, and in anticipation of obtaining programme sponsor's board approval, the programme sponsor needs to develop proposals for executing Stage C:

- Develop the terms of reference for Stage C

- Develop, in conjunction with the programme manager, a time schedule for Stage C

- Develop, in conjunction with the programme manager, a resources plan for Stage C

- Secure programme sponsor's board approval to proceed to Stage C

3.3.2 Programme business change manager

During Stage B, the programme business change manager has responsibility for the following:

- Supporting the programme sponsor in the production of the programme brief

- Verifying that the information contained in the programme brief reflects the business objectives of the sponsoring client body

- Defining, based on the objectives set out in the programme mandate, the characteristics and nature of the benefits to be delivered

- Supporting the programme sponsor in the production of the programme business case

- Ensuring the benefits to be delivered by the programme are clearly stated in the programme business case

3.3.3 Programme manager

The programme manager is introduced into the programme during this stage to ensure that the information regarding the implementation of the programme is realistic and appropriate. This is a senior appointment and will require a person with high levels of leadership and a proven ability in the successful delivery of programmes. The programme manager's tasks include the following:

- Supporting the programme sponsor in the production of the programme brief
- Supporting the programme sponsor in the production of the programme business case
- Developing a strategy for the implementation of the programme
- Establishing the deliverables required to achieve the programme's benefits
- Considering an initial listing of likely projects required to achieve the identified deliverables
- Developing a time schedule for programme delivery
- Developing a cost plan for programme delivery
- Developing a risk register for programme delivery
- Developing a resource plan for programme delivery
- Developing, in conjunction with the programme sponsor, the terms of reference, time schedule and resource plan for Stage C

3 Stage B: Initiation

4 Stage C: Definition

4.1 Purpose of this stage

The purpose of this stage is to provide a detailed definition of how the programme is to be established, determine the delivery strategy, calculate the level of resources required and develop the governance controls. All of this information is contained within the programme delivery plan.

Please refer to the programme delivery matrix (PDM) in Appendix A to understand the key roles and artefacts that comprise the CIOB programme life cycle.

4.2 Key activities at this stage

Approval of the programme business case authorises the appointment of the principal members of the programme management team, who will then develop the strategy and methodology for the execution of the programme, collectively documented in the programme delivery plan (see Figure 4.1).

During this stage, concurrent with the programme delivery plan being developed, work is undertaken to develop processes and procedures for the implementation of the programme.

They provide all the programme participants with a consistent understanding of how the programme needs to be managed and delivered.

4.2.1 Benefits profiles

The vision statement and programme mandate (produced during Stage A) are reviewed to develop a greater understanding of each benefit. Benefit profiles are developed, which consider the following:

- A description of the benefit
- The way it will be realised
- How and when will this be achieved or measured?
- What measurement mechanism is appropriate?
- Dependency on, or relationship with, any other benefit or activity

The benefit profiles are collated into a benefits realisation plan, which describes how and when benefits resulting from the programme will be obtained.

4.2.2 Scope definition

Preparation of the benefit profiles will enable the development of a project list known as the 'projects register', which will be necessary to achieve the programme

Code of Practice for Programme Management in the Built Environment, Second Edition. The Chartered Institute of Building.
© 2024 John Wiley & Sons Ltd. Published 2024 by John Wiley & Sons Ltd.

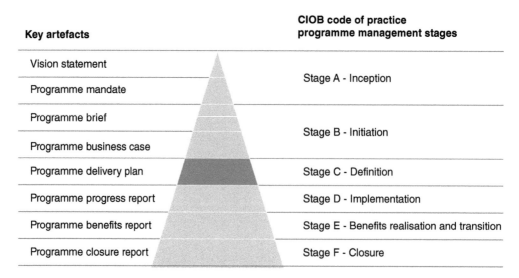

Key artefacts	CIOB code of practice programme management stages
Vision statement	Stage A - Inception
Programme mandate	
Programme brief	Stage B - Initiation
Programme business case	
Programme delivery plan	Stage C - Definition
Programme progress report	Stage D - Implementation
Programme benefits report	Stage E - Benefits realisation and transition
Programme closure report	Stage F - Closure

Figure 4.1 Key Artefacts: Stage C (Definition).

outcomes. Each identified project needs to be fully scoped, together with the criteria of output, cost and time established. Any across-project dependencies or external constraints need to be highlighted. Delivery strategies for the projects must be stated, along with the managerial resources required and an indication of where these resources will be sourced.

4.2.3 Scope of undertaking

Scope definition is among the key critical success factors often cited by senior managers. These include:

- Scope clarity and innovative design to create value and save time in delivery
- Culture and diversity exemplified by a programme team
- Strong management processes and accountability

Scope definition is fundamental to an effective programme control function in order to build the right organisation and set of policies, procedures and processes for delivery. It will guide the programme to:

- Establish a baseline for future reference and realistic standards
- Monitor performance
- Keep the plan under constant review and take action when necessary to correct the situation

It provides a comprehensive summary of a programme, detailing the scope, schedule, budget and risk based on a clear map of the work and how it is broken down in time and space using the work breakdown structure (WBS). A WBS is a hierarchical and incremental subdivision of elements necessary to achieve the end objective.

A collectively developed, shared and understood WBS is at the heart of programme success.

Also needed is a network of activity that clearly sets out the overall strategy for breaking down the programme and can also be supported by a visual schematic geographical representation to show:

- Milestone and key dates
- Discrete stages of the programme and projects, including any overlap of stages

- Sub-projects, work streams such as information technology or infrastructure
- Principal or summary activities
- Inter-relationships between the activities and sub-projects
- The objective to be achieved at each stage

4.2.4 Stakeholder analysis

As programmes tend to be large and are generators of significant change, their influence has an impact on a wider scale than just one-off projects. Because of their nature, it is also likely that programmes have a higher visibility than projects. It is therefore essential that there be a detailed deliberation of the identification, interest and influence of stakeholders.

The large impact of some programmes means there are likely to be higher degrees of opposition or dissatisfaction. As effective stakeholder analysis, management and engagement are critical to programmes, a dedicated stakeholder manager, or even a team, may be assigned to ensure this important aspect is dealt with effectively.

A detailed stakeholder analysis and a stakeholder management strategy will be developed from the initial considerations included in the programme business case. A programme communication plan will be created that defines how each stakeholder will be dealt with and the nature, extent and frequency of communications within and outside of the programme team and stakeholders.

4.2.5 Risk, opportunities, issues, assumptions and constraints

The vision statement and programme mandate (produced during Stage A) will have identified the initial risks, opportunities, issues, assumptions and constraints, and these are now revisited in light of further understanding of the programme. This is an opportunity to take advantage of a wider consultation on these aspects.

Registers must be maintained of risks, opportunities, assumptions and constraints. They may be individual or combined depending on how the programme team feels they could be managed effectively.

Risks (circumstances that may occur or change, some identified and expected, others unknown and unexpected) will be analysed in a risk register, and a methodology will be adopted for their ongoing monitoring and management.

The risk register can be a qualitative one to allow the initial capture of the key risks facing the programme. Depending on the risk exposure, further studies may be undertaken to produce a quantitative risk register for the programme.

Some financial contingency allowance needs to be calculated for the impact of the risks. This activity can also include the identification of future opportunities that may be incorporated into the programme or projects.

Similarly, issues (things that have already happened and need action) will be listed in an issue register and a methodology adopted for their ongoing monitoring and management.

Initial assumptions need to be reviewed to clarify likely impacts on the programme. Some assumptions may concern factors outside the direct influence of the programme, and these may need to be referred to the programme sponsor, who may in turn need to obtain further advice from a higher level in the client organisation.

Any additional assumptions made during this stage should also be considered and included in the register. Where appropriate action plans are developed, a methodology for regularly reviewing assumptions should also be developed.

Any constraints need to be subjected to a similar approach. Consideration should be given to devising strategies that avoid major constraints. Constraints may be created by:

- Operational requirements
- Funding issues
- Commercial and political sensitivity
- Regulatory and legal requirements
- Timing issues to do with availability (legal agreements, ownership, resources)
- Key calendar dates

4.2.6 Programme timescale plan

Based on the information contained in the programme brief, the benefit profiles, the projects register and the risks, issues, assumption and constraints registers, an overall delivery time schedule for the programme is produced. (Note: to avoid confusion between terms used to describe Gantt charts, 'time schedule' is adopted throughout this publication rather than 'plan' or 'programme'.)

The time schedule estimates the anticipated completion of the programme, the anticipated duration and timing of individual projects, and identifies when benefits realisation is expected. It highlights dependencies between projects and any external dependencies. In situations where there is a complex pattern of dependencies, it may be helpful to produce dependency charts.

It should be possible to identify the critical path sequence through the programme time schedule, as this is useful information when having to make decisions on adjusting priorities. It should be acknowledged that, at this stage, it may not be possible to identify all the projects that are required to achieve the programme's objectives.

In addition to the overall programme time schedule, a more detailed start-up time schedule should be produced describing the timing of activities necessary to mobilise the next stage of the programme once approval to proceed has been obtained.

4.2.7 Programme controls

It is important that a suitable framework be developed as early as possible to monitor and control all the key inputs and outputs of all the efforts undertaken by the programme team (see Figure 4.2).

4.2.8 Programme financial plan

At the point at which the delivery plan and delivery time schedule have been devised, it is possible to prepare a financial plan that collects all the costs that have been identified in relation to implementing the programme and that sets the programme budget.

The financial plan should be compared with the financial information contained in the programme business case and any significant variances highlighted.

From the financial plan and the delivery time schedule, the expenditure cash flow profile can be calculated. This is a critical document because it informs the funding commitments and their timings for the programme. It must be presented, discussed and

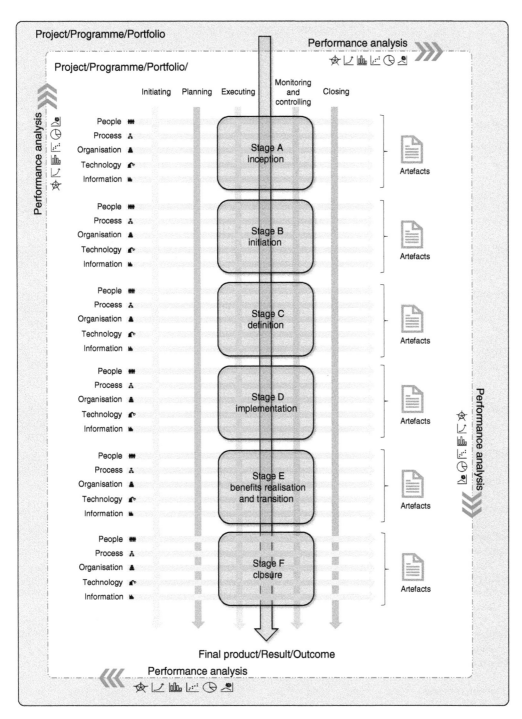

Figure 4.2 Programme controls.

approved by a senior executive of the client organisation, before it can be included in the programme delivery plan.

Compared with a project budget, there is a higher degree of uncertainty attached to determining programme budgets, as the timescales involved are likely to be much longer and the exact number and scope of projects to be instructed may not be known until later in the implementation of the programme.

The high cost of many programmes creates a substantial risk to the financial standing of the client organisation; therefore, oversight of the financial aspects of programmes is a crucial function.

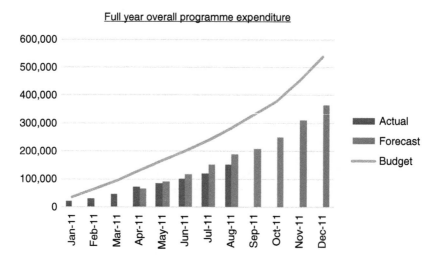

| Profit and loss | Month | Year to date | | | | Year end |
August 2022	Actual	Actual	Forecast	Budget	Variance forecast	variance budget
Turnover						
EBITDA						
Depreciation						
EBIT						

| Financial results | Month | Year to date | | | | Year end |
	Actual	Actual	Forecast	Budget	Variance forecast	variance budget
Profit before tax						
Tax						
Profit after tax						

Balance sheet	Aug-21	Aug-22	Movement
Fixed assets			
Tangible assets			
Current assets			
Short term loan			
Cash and cash equivalent			
Liabilities			
Operating liabilities			
Net current assets (liabilities)			
Capital and reserves			
Share capital			
Share premium			
Retained earnings			

Figure 4.3 Full-year programme expenditure example.

The role of programme financial manager must be undertaken by an experienced financial manager who has expertise in dealing with complex issues such as tax liability, capital allowances and programme funding (see Figure 4.3).

4.2.9 Transition plan

Prepared principally by the business change manager, a detailed strategy must be produced to describe the mechanism by which each project output is handed over

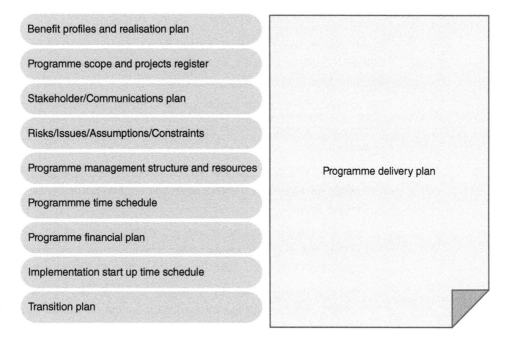

Figure 4.4 Programme delivery plan.

and incorporated into the enterprise that will maintain and operate the outputs. How this is achieved will vary depending on the nature of the undertaking.

The transition plan will explain what management infrastructure must be in place for the enterprise to take ownership of project outputs as they become available. The document should also set out the acceptance criteria for each of the deliverables.

4.2.10 Programme delivery plan

Production of the documents discussed collectively forms what is referred to as the programme delivery plan. The programme delivery plan is a detailed description of what the programme will deliver, how and when it will be achieved and the financial implications of its delivery.

The programme delivery plan (see Figure 4.4) is presented to the programme sponsor's board for their consideration, and provided it falls within the parameters established in the business case, their endorsement that the programme can proceed to implementation is given.

In situations in which the client organisation either has a complex structure or comprises a number of separate legal entities, or where stakeholders have a critical influence on or are significantly affected by, the programme, the programme sponsor's board may require further consultation on, or consent to, the programme delivery plan, prior to their authorisation.

4.2.11 Establishing the programme organisation

- Appoint the remainder of the programme management office and any other members of the programme management structure

- Arrange for the establishment of a physical location for the programme team

- Establish the information technology infrastructure to support the programme

- Set up the programme-wide systems, such as intranet sites and information management

4.2.12 Benefits management

At this phase, it is necessary to define how benefits will be managed, from identification through realisation, in alignment with the programme vision statement and the business case.

The key considerations at this phase will include the following:

- How will the benefits be identified – for example, through benefits mapping workshops

- Who needs to participate – typically programme manager, programme business change manager and relevant stakeholders and participants from the programme delivery team

- How will the benefits be monitored/reviewed and when – it is important that benefits are regularly reviewed as programme delivery progresses to allow for requirement changes, risks, issues and changes to the initial assumptions

- What are the key criteria for reviewing – for example, is there focus on the right benefits, can the targeted value of each benefit be improved, can the costs associated with each benefit be reduced, are there any additional benefits that can be targeted, are the risks and issues associated with benefits being dealt with

The considerations and outputs generated at this phase must be subject to a change control process, and benefit profiles can be used to capture any changes or amendments. The benefit realisation activities in the programme delivery plan will also need to be amended to reflect any approved changes.

4.2.13 Change management

- Manage the process of appraising changes as instructed by the programme sponsor

- Advise the programme sponsor and programme sponsor's board of the implications for the programme and its deliverables of the changes incorporated

4.2.14 Assurance and audits

- Depending on the nature and complexity of the programme, it may be necessary for the programme sponsor and/or programme manager to initiate either internal or external audits of the programme or projects

4.2.15 Transition

- The programme business change manager incorporates project outputs into the business operations of the new enterprise in accordance with the transition plan

4.2.16 Stakeholder analysis

An effective way to manage stakeholders is to divide them into common groups. Common stakeholder groups include the following:

- General public (people who are only indirectly affected by a programme but who may have a significant influence on its realisation)

- Community (people who are directly affected by a programme through their geographic proximity to programme works)

- Employees (delivered changes and benefits will directly impact employees of the client organisation)

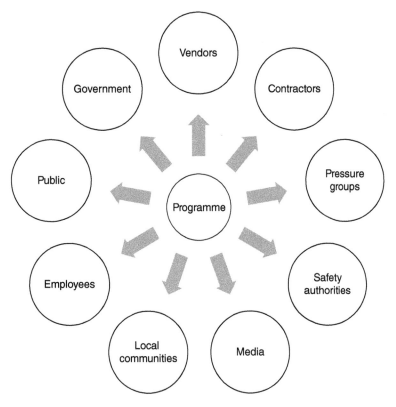

Figure 4.5 Stakeholder map.

- Shareholders (individuals or legal entities owning shares in the client organisation that is undergoing change who are affected by the business change)

- Customers (client organisations whose customers will be affected by the business change)

- Interest groups (members who share common interests and control some area of activity, such as non-profit organisations and volunteer organisations)

4.2.17 External environment and relationships

Any programme affects the environment that it operates in and, at the same time, is also affected by the environment. The external environment influences business strategy and the success of a programme.

It is therefore essential not only to identify external drivers in order to shape a strategic change and a programme but also to identify, map and track programme stakeholders, as this will help develop support and manage programme communications (see Figure 4.5).

Programmes and organisations need to be able to identify their stakeholders and judge the level of power they hold to affect the decisions and outcomes of the programme. The first step in this process will be to create a stakeholder map. This map will include all the stakeholders for their organisation with the programme at the centre (see Figure 4.6).

4.2.18 Scope management

The scope of a programme relates to the collective outcomes/outputs of all the component projects and activities. The responsibility for managing individual project

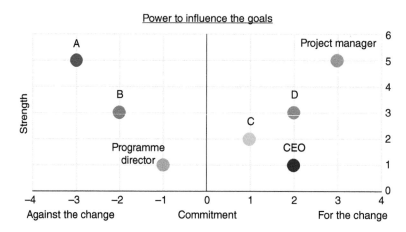

Figure 4.6 Ability to impact and commitment to the change.

scopes will stay with the project managers; however, the programme manager will be responsible for the overall programme scope.

The scope of a programme is underpinned by three key components:

- Scoping of information from initiation documentation as prepared in the earlier phases, such as the vision, mandate, brief and business case

- Definition of requirements, which is effectively a detailed design of programme deliverables

- Delivery of requirements, which continues from the programme design and defines the scope of work for each activity or project

The delivery requirements collectively provide a view of the end state of the programme and must align with acceptance criteria to ensure that the project activities address and satisfy the specific requirements.

The programme initiation documentation outlines the current state and the vision for the future state; it is the scope that defines how the programme will enable the changes to ensure that the transition is affected from the current state to the desired future state.

Where the end state is well understood and has a tangible output (e.g. in construction and engineering), it is usual to define the scope as accurately as possible at the beginning. This potentially reduces the level of changes that may be required and keeps costs from escalating. It is also necessary to define what is outside of the scope to avoid misunderstandings. Clearly defining what is in and out of scope reduces risk and manages the expectations of all key stakeholders.

Where the objective is less tangible or subject to significant change, for example, business change or some information technology systems, a more flexible approach to scope is needed. This requires a careful approach to avoid escalating costs.

It is important to recognise that, particularly for large and complex programmes, it is most likely that initial scoping will be required to undergo changes as risks, issues and changes in the wider landscape emerge. The programme design must enable a robust and effective change management process to deal with the scope variations and changes, and most importantly, it must be capable of identifying and flagging where a change in scope occurs.

It is the responsibility of the programme manager to flag any changes to the scope (be they at project level or programme level) to the programme sponsor and all changes to scope must be authorised by the programme sponsor's board.

In practice, the majority of scope change requests will be generated at the project level; it is critical that scope change approval is not done at the project level. Those involved in projects can see their own work, but they cannot see the interdependencies that exist between projects and which may impact the programme. Therefore, those working on the projects do not have the right level of understanding as to the impact of scope change requests across the projects. This requires that project scope changes be escalated to the programme level for a decision through formal governance controls.

Scope is typically managed both at programme and project level.

- Programme scope: This is owned by the programme manager, and changes to programme scope are managed at the programme level. This may lead to changes necessary at project levels as well.

- Project scope: The programme delivery plan will identify the processes required at the project level in terms of managing scope. Project scope changes must be examined at the programme level to ensure they are monitored and actively managed. These project scope changes must be elevated to the programme for approval. The impact of an approved scope change request is communicated to the affected projects for management.

Managing programme and project scope change is one of the primary responsibilities of the programme management office.

These methods are used to identify the activities that are necessary to achieve the programme's objectives:

- Work breakdown structure

- Scope statement

- Acceptance criteria

- Exclusions from scope

- Assumptions

4.2.19 Risk management

Risks will be identified, recorded, monitored and managed in the following ways:

- Risk management methodology

- Risk register

- Risk review process

Risks, issues and opportunities

As technical and commercial issues get more complex and financial metrics tend to proliferate, finance and programmes often end up measuring performance and risks in different ways, using various sources of information and, in many cases, using a different language. Risk assessment is not an exact science, and there are a number of different methods to measure or quantify risk.

Effective risk management begins with realism – seeing things as they are – and continues with a joined approach and common understanding at all levels of an organisation or programme. A strong corporate risk management culture and consistent risk-rating methodology are fundamental to the success of a programme in order to focus on the risks that matter.

The guidance below offers a structured, common sense-based approach to risk management.

Success factors

The key success factors are as follows:

- Identification and control of risk
- Alignment to business value drivers
- Awareness of changing risk profile and risk appetite
- Comprehensive approach to risk management

Risk categories

Risks will generally fall under the following categories:

- Financial risks
- Reputational risks
- Health and safety risks
- Operational risks

Objectives

The objectives of effective programme risk management are as follows:

- Provide a mechanism for the early identification and resolution of risks that may arise
- Ensure that risks are escalated to and mitigated by appropriate levels of management
- Provide the visibility of risks that may affect or are affecting high-priority projects
- Provide accountability for the mitigation of project risks
- Provide guidance for the correct control and administration of the recorded risks
- Provide a basis for determining the level of financial contingency required for the programme

Definitions

Risk is often defined as 'an effect of uncertainty on the objectives'. Risk can be categorised as the following:

- Hazard – risk of adverse events
- Uncertain outcome – not meeting expectations
- Opportunity – exploiting the upside

An issue may be defined as:

- Any unresolved problem that could jeopardise programme outcomes
- A risk that has materialised
- An uncertainty, which was not previously raised on the risk register, has occurred

It is imperative that all issues are recorded on the project issue log; however, an issue should be reported to the programme management office only if the project team cannot resolve the issue or if the project manager determines that the issue impacts other projects that are outside the PMO's scope of responsibility. The issue process implemented by the programme management office is to ensure that the appropriate management team is engaged to help resolve problems.

At the portfolio level, risk management and change control are supported centrally, which brings together a set of essential functions to support the successful delivery of programmes and projects, including:

- Coherent upward reporting to the management board on its key programmes and projects to support effective decisions

- Timely sharing of information and lessons learned through outward relationships ... and beyond

- Inward support to help deliver programmes and projects with the right expertise when needed

Quantitative risk assessments based on experiential understanding of the effect of uncertainty in relation to the programme out-turn cost (or anticipated final cost) and schedule delivery will be performed at key stages of a programme. The certainty of outcome will increase as the programme approaches its final stages. Probability management will store uncertainties as data in a model that is actionable, additive and auditable.

For this purpose, three-point estimating can be used as a tool for estimating cost and schedule value and assessing overall risk. See Appendix B for further details.

4.2.20 Reference class forecasting

Most estimating techniques use a ground-up approach. There is another technique that uses a top-down and external view. Developed on the thinking of Daniel Kahneman and Amos Tvorsky, reference class forecasting is a great tool for assessing programme performance (Kahneman, 1977).

Reference class forecasting accommodates not only the known unknowns but also the unknown unknowns (Budzier, 2018).

Looking at a range of similar completed events can provide a better estimate and a reference check against estimates built up in other ways. Reference class forecasting provides an outside view of a forecast (Flyvbjerg, 2004).

The steps to producing a reference class forecast are as follows:

1. Identification of a relevant reference class of past similar projects

2. Establishing a probability distribution for the selected reference class

3. Comparing the specific project with the reference class distribution, to establish the most likely outcome for the specific project

As an example, Figure 4.7 shows the frequency distribution of past projects, in particular the percentage cost overrun. This information can be plotted, and a good prediction of future performance can be ascertained from past data.

The correct use of reference class forecasting can reduce optimism bias and strategic misrepresentation of programme estimates (Fyyvbjerq, 2016). The UK government's Green Book defines optimism bias as the demonstrated systematic tendency for

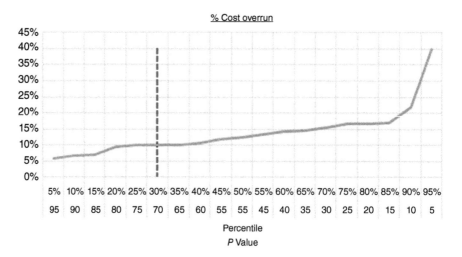

Figure 4.7 Reference Class Forecast.

appraisers to be over-optimistic about key project parameters, including capital costs, operating costs, project duration and benefits delivery (HM Treasury, 2022).

The Department for Transport in the UK also recognises reference class forecasting as a suitable method for reducing optimism bias in transport planning (Department for Transport, 2020).

Appendix B contains further information on how to perform reference class forecast calculations.

4.2.21 Programme schedule and risks

The likely duration of a schedule risk can be ascertained by using quantitative analysis, and the results can be plotted for review.

Figure 4.8 shows an example of where the risk durations are shown for a programme consisting of five separate projects. Reference class forecasting can be used for managing multiple projects within programmes.[1]

4.2.22 Programme governance

Governance defines the structure, roles and responsibilities to set objectives and report and monitor performance in order to make decisions and steer a programme towards its anticipated destination.

Effective governance of programme management ensures that an organisation's programme is aligned with the organisation's objectives, is delivered efficiently and is sustainable. The governance of programme management also supports the means by which the board, and other major project stakeholders, are provided with timely, relevant and reliable information.

The governance of programmes will cover the following areas:

- Leadership and sponsorship, clear and documented roles and responsibilities

- Strategic direction: a multidimensional roadmap to reach the final destination

- Programme management methodology, including a set of policies, procedures and processes to control programme trajectory against baseline

---

[1] Hanif & Aldhaheri, Managing multiple projects using reference class forecasting, INDIS, Novi Sad, 2021

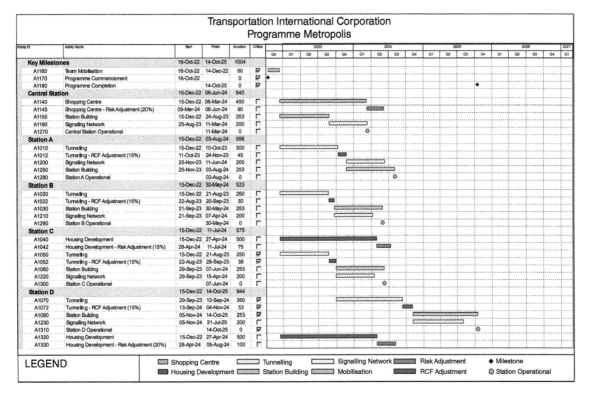

Figure 4.8 Programme schedule with risks.

- Integrated assurance – transparency and disclosure to all stakeholders via operational (projects), functional (PMO, department) and compliance (risk and audit) lines of defence

The terms of reference for a programme board will typically cover the following:

- Requirement – why?

- Role – what?

- Composition – who?

- Conduct – when?

- Proposed delegated authority – how/how much? This is based on a funding agreement and contingency allocation.

Optimal governance, programme change control, risk management and reporting of an approved baseline scope will allow a business to manage risk, deliver value and drive programme and project decision-making while saving time and effort (see Figure 4.9).

Establishing and implementing a measurement and reporting system is a complex and evolving process based on the simple business principle that what can be measured can be managed. 'Too much data, not enough information' is a common complaint among business managers who have to keep track of the status of programmes while relying on information that is incomplete, out of date, inaccurate, late or simply irrelevant.

Reporting requirements will change as projects and programmes move from inception through their life cycles. It is important that the reporting system is designed to focus on metrics that matter at the relevant stage.

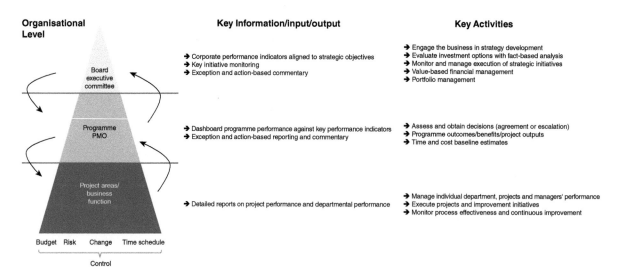

Figure 4.9 Change, risk and reporting.

The critical success factors that underpin the operation of effective reporting are the following:

- Strategic objectives – Performance measures need to be aligned with the strategic objectives of the organisation. These objectives must be clearly communicated and understood by employees and external stakeholders.

- Baseline management – The baseline scope, programme timescales, resources and budget cost for the approved brief are clearly and consistently communicated to all parties.

- Work breakdown structure – A WBS is designed to provide a common basis for linking the scope of work, estimates, budgets, schedules, earned value progress, performance and cost reports, based specifically on the customer's brief.

- Progress management – As the programme progresses, the actual delivery times, resources and costs are recorded, analysed and compared to the baseline.

- Change management – All changes from the original baseline are managed and incorporated into subsequent controlled copies of the original baseline called 'the current baseline'. The current baseline is progressed to reflect the current situation and is subjected to analysis.

- Reporting – All baseline and progress information is collected, analysed and reported in a simple yet robust manner based around schedule, cost and quality performance reports, clearly communicating the status of the delivery brief to all parties, including the client.

The measures that matter for effective control should be simple and reliable.

Key performance indicator themes should be reported and agreed upon by each functional area or, in leading global organisations, across a portfolio of projects or a programme to deliver strategic objectives. The indicators are then aggregated into project, programme and executive board reports to assist in meeting the needs for information, control and governance. As the programme progresses through its life cycle, the quality and availability of key performance data available to the different functions will develop.

Performance measures should be based on the key performance indicators that are most commonly associated with the built environment and include key objectives such as cost, time and quality.

Organisations will also seek to measure and report other success factors that align with their strategic objectives, such as environmental sustainability and corporate social responsibility themes. The importance of these performance indicators and measures will change as the capital projects and programmes move through the various stages of their life cycle (investment planning, design, procurement, manufacturing, construction, commissioning and operations).

Once this process is in place and data and information are flowing more or less painlessly, it is then up to management to trust it and act on it, as there is often no worse decision than no decision at all – even a bad one.

This trust will undoubtedly be reinforced by a level of checks at all levels of the organisation, or integrated assurance.

4.2.23 Issues management

The programme's approach to issue management must be clearly defined and endorsed by the programme sponsor, including the decision-making procedures.

The approach should include roles and responsibilities in relation to management of issues during the programme and may include the following:

- How will issues be flagged, passed up the hierarchy, recorded and assessed?

- How will issues be monitored and controlled (both at project level and programme level)?

- Who will generate the report and when, and who will receive the reports and determine action?

- What will be the regular review points?

- What will be the escalation procedures?

Initial issues will have previously been recorded as part of the completion of the brief and the business case – the initial registers can be turned into the live issues register with assigned ownerships as the programme progresses.

4.2.24 Time scheduling

In its simplest form, a schedule is a listing of activities and events organised by time. In its more complex form, the process examines all programme activities and their relationships (interdependencies) to each other in terms of realistic constraints of time, budget and people, that is, resources. In programme management practice, the schedule is a powerful planning, control and communications tool that, when properly executed, supports time and cost estimates, opens communication among personnel involved in programme activities and establishes a commitment to programme activities.

However, it is not always practical or realistic to prepare a time schedule for programmes – in many instances, there may be additional constraints (e.g. output delivery dates may be set by parameters beyond the control of the delivery team or a target may be set for commencement of benefits realisation for certain outcomes); it is often the case that, especially for programmes with intangible deliverables, while a realistic time schedule can be prepared for individual projects and activities, the overall programme time scheduling remains a bit of an unknown until the commencement of the implementation phase. However, almost all business cases will allocate a certain time for any programme, even if only for funding purposes, and it is the responsibility of the programme manager to ensure that a detailed programme schedule is prepared at the definition phase, identifying each individual component, and that it is regularly reviewed for any variance. The programme sponsor's board will be made aware of the variances through the programme highlight report and should decide on the appropriate course of action.

4.2.25 Financial management

Business today needs finance professionals who understand how construction and production processes are really managed to help with establishing:

● clarity in budget and contingency allocation between funder, programmes and projects

● understanding of reporting requirements at corporate and programme level in order to prevent misalignment, inconsistencies and misinterpretation

Difficulties in aligning and reporting for business and programme purposes are not uncommon. Financial reporting normally spans a fiscal year, whereas programmes, due to their nature, may span a number of years. Additionally, financial and programme professionals do not necessarily speak the same language or have the same reporting needs.

Communication is the key aspect of prudent financial management. Opacity in budget allocation leads managers to make assumptions and decide on the basis of their current knowledge over time. Particularly on major programmes, it is of vital importance that meaningful and timely information is generated from the vast amount of data that would be available. Figure 4.10 sets out the individual roles of the key personnel in the context of programme management at this stage.

Funding arrangements

Once funding arrangements, share of contribution, delegated authority and contingency allocation are apportioned and agreed upon to form a fund, a budget developed in parallel can be managed and its change controlled through the programme gates.

Programme Manager	Programme Finance Manager	Project Manager(s)
● Ensure that appropriate financial controls are defined and operate for the programme ● Ensure that programme operates within the allocated budget ● Provide timely reports to programme sponsor and programme sponsor's board of any significant variances from the approved budget or changes and or issues that may result in a variance	● Provide advice and guidance on the proper financial management of the programme and each project or activity ● Oversee compliance and governance in terms of financial management requirements ● Maintain up-to-date information on budgets, forecasts and expenditures ● Provide regular reports to programme manager, programme sponsor and programme sponsor's board	● Ensure that any overarching financial management and governance requirements and enshrined within project documentation ● Ensure that projects are delivered within the allocated budget ● Provide regular reports to programme manager and project board

Figure 4.10 Financial management roles and responsibilities.

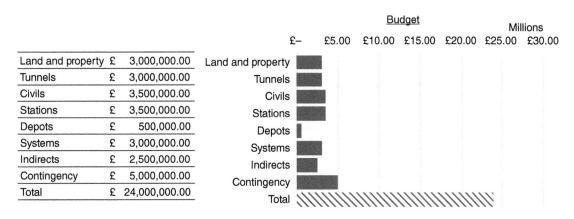

Land and property	£	3,000,000.00
Tunnels	£	3,000,000.00
Civils	£	3,500,000.00
Stations	£	3,500,000.00
Depots	£	500,000.00
Systems	£	3,000,000.00
Indirects	£	2,500,000.00
Contingency	£	5,000,000.00
Total	£	24,000,000.00

Figure 4.11 Programme budget for transport programme (example).

Budget, including contingency, can be represented by a WBS element using a tabular format or a bar chart, as seen in Figure 4.11.

It is essential to capture baseline assumptions as part of a budget to allow all parties to understand the associated risks. Assumptions will typically include scope, design, construction, schedule, cost, inflation, currency, commercial, procurement, land and property, legal framework, operations and asset management.

Changes and contingency management related to scope (schedule, cost and quality), including design development, valuation (value engineering), new scope, sequencing or acceleration cost (also known as de-risking in some programmes), cost savings or procurement and delivery model adoption (e.g. engineering procurement and construction as opposed to in-house delivery), can be identified and communicated using a waterfall or bridge diagram.

Waterfall charts are commonly used in business to show how a value changes from one state to another through a series of intermediate changes. Waterfall charts are often called bridge charts (particularly in financial jargon) because a waterfall chart (Figure 4.12) shows a bridge connecting its endpoints. In addition to this, Sankey charts can also be used to depict the flow of funds and help in understanding the size of the various components of the programme (see Figure 4.13).

4.2.26 Cost management

Traditional cost and performance measurement systems often fail to distinguish between cost incurred and physical progress made. Under these systems, it is difficult to analyse separately cost variances and schedule variances. One answer to this is earned value management.

There are two key primary objectives of an earned value system:

i. to provide programme project managers with a reporting system that gives them better control over cost and schedule management and

ii. to provide customers with a better picture of the status of work. See Appendix for a monthly programme report template.

Cost performance indicators (CPIs) and schedule performance indicators (SPIs) are commonly used in the construction industry. CPI = 1 and SPI = 1 means the programme is on track.

CPI and SPI are the two principal components of earned value management (see Figure 4.14) and allow project and programme to:

		Values
Start	5,000,000	5,000,000
Jan	-	250,000
Feb	-	200,000
Mar	-	60,000
Apr	-	180,000
May	-	200,000
Jun	-	500,000
Jul	-	100,000
Aug		100,000
Sep	-	300,000
Oct		75,000
Nov	-	300,000
Dec		50,000
End	3,135,000	

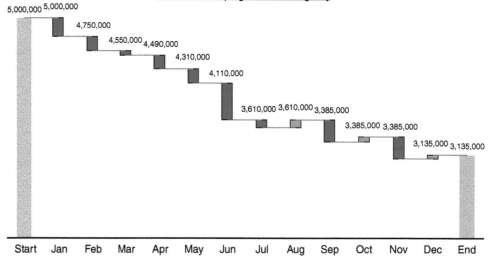

Figure 4.12 Waterfall chart (programme funds example).

- Measure programme and project progress
- Forecast completion date and final cost
- Identify schedule and budget variances along the way

There are several secondary objectives for measuring CPI and SPI on projects across a major programme. These include:

- Providing a timely 'early warning' signal for prompt corrective action
- Comparing the amount of work performed during a period of time indicates whether the project is behind or ahead of schedule
- Comparing the budgeted cost of work performed (BCWP) with actual cost indicates whether the programme is over budget or under budget
- Encouraging contractors to use effective internal cost and schedule management systems

By integrating these measurements, earned value management aims to provide consistent indicators to evaluate and compare the progress of programmes and construction projects. Earned value management compares the planned amount of work with what has actually been completed to determine if cost, schedule and

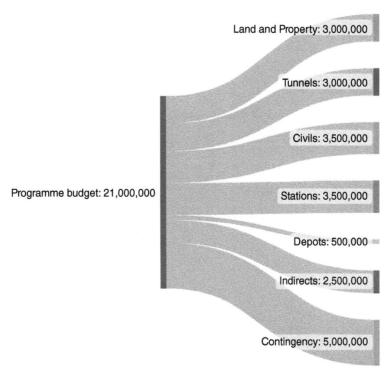

Figure 4.13 Sankey chart (programme funds example).

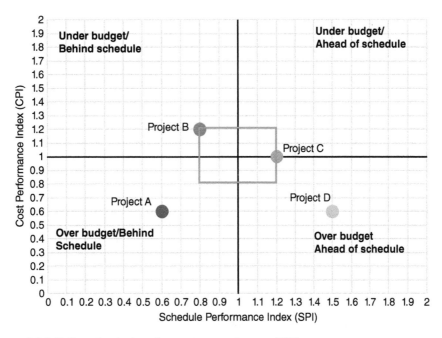

Figure 4.14 Delivery/project performance – programme EVM summary.

work accomplished are progressing as planned. Work is 'earned' or credited as it is completed.

CPI and SPI are not absolute measures of progress and should not be viewed in isolation. The indicators should be viewed in conjunction with other progress reports and performance measures, such as monthly milestone and variance reports. The information is processed and published in progress reports on a monthly basis and, in conjunction with other project progress information, can be used by both the client

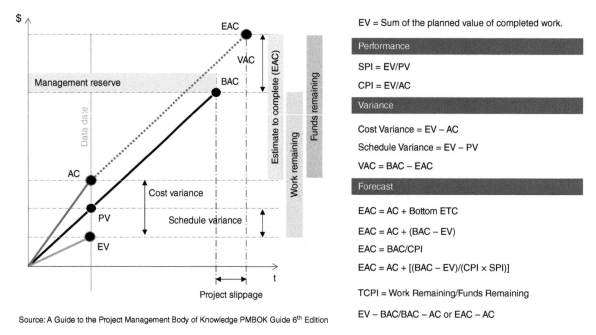

EV = Sum of the planned value of completed work.

Performance

SPI = EV/PV

CPI = EV/AC

Variance

Cost Variance = EV – AC

Schedule Variance = EV – PV

VAC = BAC – EAC

Forecast

EAC = AC + Bottom ETC

EAC = AC + (BAC – EV)

EAC = BAC/CPI

EAC = AC + [(BAC – EV)/(CPI × SPI)]

TCPI = Work Remaining/Funds Remaining

EV – BAC/BAC – AC or EAC – AC

Source: A Guide to the Project Management Body of Knowledge PMBOK Guide 6th Edition

Figure 4.15 Earned value metrics.

and supplier(s) to monitor performance and encourage a greater degree of control and proactive decision-making.

To make a positive contribution, the measurement system is dependent on two key factors:

- Implementation and integration of a collective and accurate breakdown of the construction works that capture the planned schedule (time) and budget costs throughout the lifetime of the project or programme. This is referred to as the cost-loaded schedule.

- Implementation of a robust and rigorous system to regularly capture actual time and cost data in order to compare it with that planned.

Other earned value metrics

CPI and SPI are the most commonly used metrics; however, there are a whole host of others that can also be used to measure performance (see Figure 4.15).

Programme forecasts

Programmes will also report an annual spend forecast (Figures 4.16 and 4.17) or fiscal year performance and provide long-term projections (Figure 4.20) based on the approved baseline. This should be aligned and coordinated with financial reporting.

Although a source of many frustrations, integration is key to programme and financial reporting alignment and will drive accurate reporting and inform decision-making.

One method is to set up the main accounting systems to track cost and billings by WBS line item. Most enterprise resource planning packages will track inventory movements by item type or by project code.

Project- and programme-oriented businesses need to be able to relate the use of material and labour to one key 'manufacture' or 'procurement' item number at a WBS level, allowing:

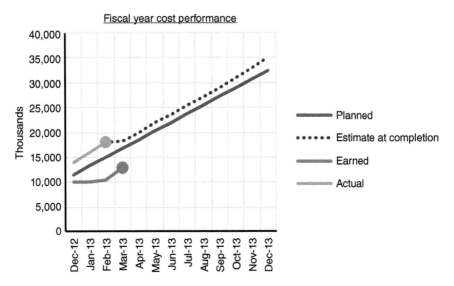

Figure 4.16 Programme fiscal year performance (annual spend forecast).

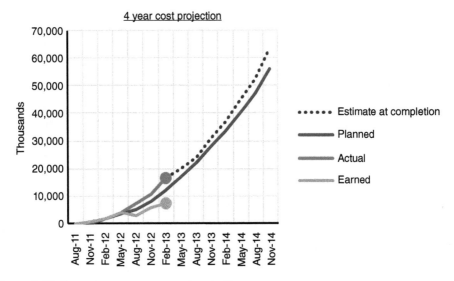

Figure 4.17 Four-year programme cost projection.

- Meaningful comparison between the estimated or BCWP and the actual cost of work performed (ACWP)
- Real-time update of the estimates or standards for costs to compete against actual costs based on incurred costs

Best practice strategies for improved cost control include:

- Cost management planning
- Supplier involvement
- Value engineering
- Cost reporting
- Programme cost change management
- Cost forecasting
- Risk reporting

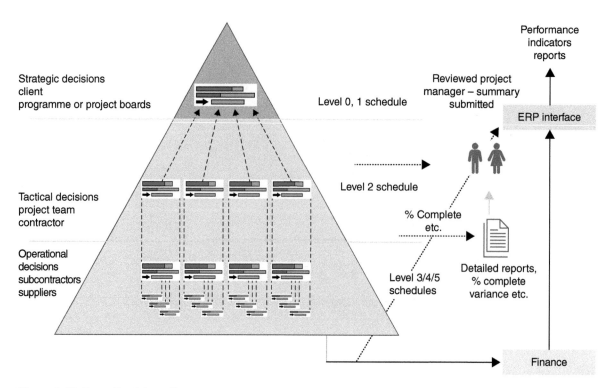

Figure 4.18 Reporting integration.

Finance will control the programme expenditure at year end and monitor actual against forecast (AFC) expenses and budget. Profit and loss and balance sheets will also be compiled for business-reporting purposes and will include programme cost to date and any other additional costs outside programme control. (See Figure 4.18).

4.2.27 Change control

At the definition phase, it is critical to ensure that a robust change control and management procedure is designed to enable formal identification, evaluation and decision-making required for accepting or rejecting changes to aspects of the programme.

Procedures should be put in place to manage changes, both at programme level and project level. Changes to scope, programme delivery plan, business case, benefits and budget will affect the overall programme and must be managed at the programme level.

Programme-level changes must be signed off by the programme sponsor's board and managed by the programme manager. Typically, programme highlight reports should contain a change register.

4.2.28 Information management

Process and systems to be used for managing and storing information include:

- Programme-level information technology systems and processes

- Responsibilities for issuing, distributing and maintaining information

- Archiving and retrieval protocols

Information management

Information has to flow within a programme in a way that will allow management, project managers and project teams to achieve business objectives while transparently keeping all stakeholders informed.

Effective communication is a two-way process in which technical information, analysis, commentary and comments are conveyed through reports, speech and other mediums to a specific audience so that the intended audience will be informed, perform an action, or reach a decision based on this information.

The fundamental purpose of information management is to communicate effectively and with consistently relevant and accurate data to project stakeholders at all stages of the programme life cycle. The provision of good quality, timely information is an essential deliverable for any programme. Systems and procedures should be defined to enable effective management of information through the life cycle of the programme from inception to the delivery of operational assets. An effective document management system should facilitate direct relationships with the relevant information, enabling storage and 'retrievability' of the following:

- Document ID

- Document location

- Document classification

- Document function

- Document status (cost and time)

- Document criticality and confidentiality

4.2.29   Integration

Integration is the combination and coordination of multiple sources of related information so that sources work together and form an integrated whole with the ability to provide a snapshot of progress status at any chosen reporting milestone. The integration process will allow for interdependencies and design, procurement, construction, commissioning, departmental, organisational and supply chain interface issues to be identified and addressed from a single source. Key information (variance, exception) will be captured, assessed and forwarded to the relevant authority for a decision to be made and implemented.

The level of integration will vary from loose to tight depending on the current programme stage, agreement of a common purpose, set of objectives and level of process and system standardisation. Programmes are usually carried out by a project team under the overall direction and supervision of a programme manager. In practice, there will be many variants of this structure, depending on the nature of the programme, the contractual arrangements, the type of project management involved (external or in-house) and the client's requirements.

As the programme evolves, it is essential to strive for an integrated information repository to manage increasingly detailed information. All integrated programme assets or documents should be coded and linked to a programme database. In the context of a physical asset programme, BIM is a powerful tool that will allow a programme to be integrated into a multidimensional model created for management purposes.

In summary, programme management systems need good quality sources of information on which to base their decision-making and manage operations post-delivery. Management reports, project summaries, reporting systems and data will include the following:

- Programme and project costs incurred to date and current estimate also known as out-turn cost (i.e. how much will it all cost in the end?)

- Cash flow status on programme and project and estimated cash flow forecast for the remaining period of programme

- Earned value measures of schedule variance and cost variances for project work packages and overall programme

- Status of risk management actions in relation to the database of programme risk registers

- Estimates of risk and uncertainty remaining in the key elements of the programme

4.2.30 Communication management

The process developed for communicating with the programme team and stakeholders may include:

- Responsibilities for generating and distributing communications

- Nature of communications to be issued

- Methods of communication to be utilised

- Stakeholder analysis

- Frequency of communications

Making the link with the communication team

In simple terms, communication is the process of transferring information and knowledge from one source to another.

Effective communication of relevant information is a two-way process in which data, evidence, analysis, commentary and feedback are conveyed through reports, speeches and other media to a targeted audience to perform an action or reach a decision based on this information.

Once the stakeholder audiences are known, it is paramount to define a set of key messages and positioning statements to address these audiences in their questions, concerns and other key requirements. A series of pre-defined communication channels should be prepared, especially in times of crises or issues. It is important that messages are continuously updated and aligned with the programme's challenges.

Communications strategy

A good communication strategy and plan will ensure that the programme audience is informed in a targeted and timely manner.

Based on the stakeholder analysis, the communication plan will list the messages and interventions to be communicated to which audience group, when, how (medium), why and from whom.

Communication mediums can be categorised as high- or low-impact and commitment-building and will include:

- One-to-one coaching or small-group workshop

- Intranet, social network, knowledge transfer

- Large group, conference

- User guides, reminders and online help

Key communication principles include the following attributes:

- Benefit led

- Messages that pinpoint the benefits that are relevant to stakeholders

- Timely, accurate, up to date, regular and consistent

- Appropriate in language and media for stakeholders and staff

- Whenever appropriate, use existing channels

- Encourage two-way dialogue and feedback

4.2.31 Quality management

The method and responsibility for assessing the fitness for purpose of the project outputs, outcomes and programme deliverables will need to be defined. This might be done by specifying the acceptance criteria and/or the performance criteria. For programmes with tangible deliverables, for example, a new facility, it is relatively easier to define the quality acceptance criteria; where programmes include intangible deliverables, it may be necessary to rely on secondary assessment methods to quantify acceptance or performance criteria.

It is advisable to introduce periodic reviews of the quality management procedures at key stages of the programme life cycle to test their adequacy and effectiveness and enable any changes or amendments as necessary.

4.2.32 Procurement and commercial management

Procurement is the process by which the resources (goods and services) required by a project are acquired. Contracts are agreements between two parties. Procurement includes the development of the procurement strategy, preparation of contracts, selection and acquisition of suppliers and management of the contracts.

The benefits of contracts and procurement are that:

- A project in which procurement is aligned with the programme deliverables is more likely to meet its objectives

- Procurement often represents a major portion of project spending and hence needs careful consideration to ensure value for money is realised

- It ensures that all parties involved in the project are legally protected

Effective procurement and contract management are core to a successful programme. The procurement policy, procedures and processes should be established to guide the procurement of a range of contracts and projects.

It is important for companies to develop purchasing policies and procedures that include the following:

- Supply chain management

- Strategic partnerships

- Market knowledge

- Cost reduction

- Sustainability

Based on the above, managers will develop a 'contract packaging' approach that will break down the entire programme of work for delivery into suitable elements for procurement. This will be refined in discussion with the potential suppliers and contractors. To meet the programme objectives, on time and within the given budget, managers will often procure a large number of contracts and projects of varying size and value.

A robust, fair and transparent approach to procuring, managing and monitoring these agreements is essential, as is carefully managing the risks associated with the over-all programme and related projects. A programme will often have to operate within the procurement framework set out by international procurement legislation and national regulations.

For major programmes, it is also recommended to give advance notice of contracts through a 'future opportunities' website – the equivalent of market building for supply chain members – and adopt e-tendering practices.

To operate effectively in the marketplace, programmes will use standard procurement documentation and contracts such as New Engineering Contract (NEC)/Fédération Internationale Des Ingénieurs-Conseils (FIDIC).

Managers will award contracts according to broader objectives, values and regula-tory frameworks. Ahead of every contract competition, programmes will develop a 'balanced scorecard' of selection and award criteria that will inform companies of the commercial and technical factors their bids will be scored against. This should include time, quality, safety and security, equality and inclusion, sustainability and legacy.

Appropriate criteria will be added and different weights will be applied to each crite-rion on a contract-by-contract basis, according to the nature of the goods, services or works being procured and best practice guidance from the Office of Government Com-merce. To avoid costly and time-consuming disputes, the programme should seek to make clear agreements with all parties while agreeing and managing contracts.

Best practice suggestions to manage the risks in this area include:

- Developing a procurement plan

- Establishing vendor selection criteria

- Having an overall buying strategy for negotiation

- Establishing quality-control procedures and a set of performance measures

Key performance indicators to measure the performance of the procurement or buying processes include:

- Unit price for comparison

- Total cost of supply

- Number of suppliers

- Average purchase order size

- Number of contracts out, placed or closed

- Value of contract placed

- Satisfaction indices

Figure 4.19 illustrates an example of progress of procurement and commercial management presented graphically:

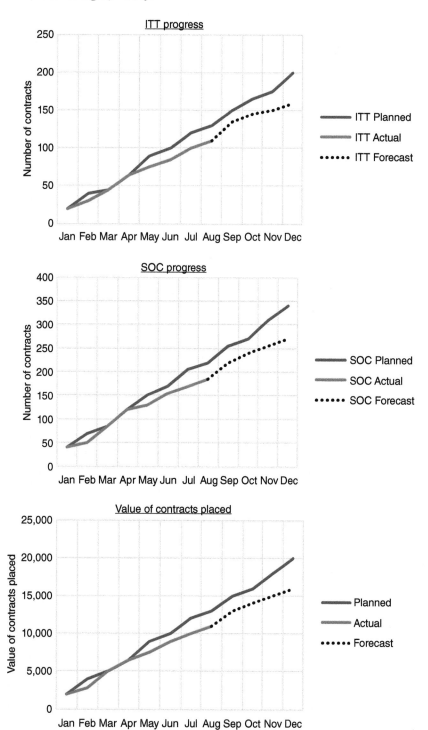

Figure 4.19 Invitation to tender (ITT), signed outline contract (SOC) plus value of contract placed.

4.2.33 Resources management

Managers need to 'design' their organisations and programmes to embrace change and deliver programmes effectively.

$$Talent \approx f(motivation, competence)$$

The following criteria will need to be considered in order to design and agree on resource profiles that are fit for their purposes and align with the delivery model so that the right level of resources are allocated to a programme at the right time:

- Work breakdown structure (WBS) – a key tool for efficient and effective resource allocation to create resource profiles for delivery

- Scope/size – work packages; scope definition, including a hierarchical list of materials, components, assemblies and sub-assemblies required to make a manufactured product

- Organisation breakdown structure (OBS) and organisational policies: structure and a clear statement of company policy on key areas such as staffing, subcontractors' issues and procurement policies

- Benchmarks and historical information. Information from past records

- Team experience and clear roles and responsibilities (using a RACI model) within a programme and down the supply chain with appropriate level of 'man marking and risk transfer'

The above criteria will depend on the delivery model adopted, but leading programme managers focus on building an intelligent client organisation to lead delivery and maintain the right level of control. This is a key factor in successful cost management. It needs to be established at the start of programme so that appropriate resources are allocated. Where necessary, certain elements may then be outsourced, subcontracted or bought according to a build-or-buy model.

4.2.34 Health and safety management

Programmes will need to formulate health and safety management policies and procedures both at the project level and the programme level. It is the responsibility of the programme manager to ensure that, at the definition phase, the health and safety goals, requirements and procedures are set and that the processes are put in place to transfer these to the project levels as necessary. In some instances, specific key performance indicators or targets may be set to measure performances both at the project level and the overall programme level. It is advisable to also set specific review stages to assess whether the health and safety management policies and procedures are performing as desired across the programme.

4.2.35 Sustainability/environmental management

It is often said that sustainability is essentially not a methodology but a dimension of thinking. The perception of sustainability is shaped by people's values, behaviour, attitude, ethos and interaction with the wider environment. Programme management is concerned more with outcomes than outputs. With emergent input in a changing environment, a programme manager makes use of current information to identify options for comparison and decision.

For sustainability in programme management, the programme manager emphasises the learning loop leading to the preferred options: the programme manager has to assess the suggested programme options in the dimensions of economic sustainability, environmental sustainability and social sustainability before recommending the options.

In most cases, the organisation commissioning the programme will have sustainability and environmental management policies at a wider level or, indeed, may have certain aspirations and targets specific to the programme. The programme manager must ensure that the goals and targets, as necessary, are defined and translated to procedural requirements at the project level.

4.3 Key roles and responsibilities of this stage

Stage C marks the commencement of the involvement of almost the full programme team as they collectively evolve the programme's implementation strategy and fully plan how the programme will be executed.

Monitoring and controlling the progress of the programme, which up to this point has been with the programme sponsor, moves to the programme manager, who now has prime responsibility for planning and designing the way the programme will proceed.

Reporting to the programme sponsor, the programme manager is assisted by a range of personnel: a programme finance manager, a programme stakeholder/ communications manager and a team of specialists within a programme management office.

The programme business change manager, working in conjunction with the programme sponsor and programme manager, continues with the function of ensuring at each stage that what is being proposed or delivered matches the requirements of the client. The business change manager takes responsibility for benefits management and for beginning to develop plans for the transition of the undertaking into its finished state.

In accordance with good management practice, the production of a roles and responsibilities matrix is a helpful device for assisting in the determination of the level of structure and resource required.

4.3.1 Programme sponsor's board

The programme sponsor's board has a number of key roles during this stage:

- Review the programme delivery plan
- Give approval to proceed to Stage D

4.3.2 Programme sponsor

Having appointed a programme manager to take responsibility for this stage, the programme sponsor adopts a more strategic role and carries out the following:

- Acts as the interface between the programme management team and the client/sponsoring organisation
- Provides direction and advice to the programme manager
- Selects and appoints, in conjunction with the programme manager, additional members of the programme management team
- Resolves any queries, ambiguities and conflicts with the programme sponsor's board
- Determines the change control process
- Review and approve the PDP

4.3.3 Business change manager

During this stage, the programme business change manager continues to focus on the benefits to be delivered and the final outcome of the programme by doing the following:

- Selecting and appointing a programme benefits realisation manager (if required)

- Developing benefit profiles to match the outcomes defined in the programme brief

- Developing transition plans

- Confirming acceptance criteria for the programme deliverables

- Reviewing and confirming, in conjunction with the programme manager, the number and nature of projects required to achieve the deliverables

4.3.4   Benefits realisation manager

Depending on the size of the programme, it may be necessary to appoint a benefits realisation manager to support the programme business change manager. The role of the benefits realisation manager during this stage is to:

- Develop the benefit profiles

- Assemble the benefits realisation plan

- Determine the mechanism for realising and measuring benefits

- Determine the critical success criteria

- Determine the acceptance criteria for programme deliverables

4.3.5   Programme manager

Leading this stage, the programme manager has a number of key tasks:

- Identifying the roles of the programme management team that need to be available during this stage

- Selecting and appointing, in conjunction with the programme sponsor, the remaining members of the programme management team required during this stage (Note: Depending on the individual circumstances, these may be internal or external appointments)

- Defining the full scope of the programme

- Arranging for a physical location for the programme team

- Arranging for the establishment of the information technology infrastructure to support the programme team

- Overseeing the production of the programme delivery plan

- Reviewing and confirming the programme's time schedule

- Reviewing and confirming the programme's cost plan

- Develop the data governance/information management strategy

- Reviewing and confirming the programme's risk analysis and risk register

- Determining, in conjunction with the programme information manager, the information/document system to be adopted

- Reviewing and confirming the programme's governance policies and procedures

- Reviewing and confirming the programme's financial policies and procedures

- Reviewing and confirming the stakeholder analysis and communication plan

- Determining the number and scope of projects necessary to achieve the programme's deliverables

4.3.6 Programme financial manager

The programme financial manager has responsibility for ensuring that the programme budget determined during this stage aligns with the business case. The financial manager's principal activities include:

- Overseeing the development of the overall programme budget

- Ensuring the budget conforms with the sums contained in the business case

- Determining expenditure cash flow forecast

- Determining the risk and contingency allowances to be included within the programme budget

- Confirming the programme's funding arrangements with the programme sponsor and programme manager

- Reviewing the cash flow forecast with the funders

- Defining the programme's financial policies and procedures

- Liaising with the financial director(s) of the sponsoring client on matters relating to the implications of the programme on their financial reporting and tax affairs

- Ensuring prudent financial governance

4.3.7 Programme management board

The programme management board comprises the senior managers of the programme management structure and provides advice and support to the programme manager. This may include senior managers from key supply chain organisations. The programme management board should meet regularly to review the programme's progress and to highlight and resolve any issues that may be hindering it.

The programme management board is likely to be composed of the following managers:

- Programme sponsor

- Programme manager

- Business change manager

- Stakeholder/communications manager

- Programme financial manager

- Head of the programme management office together with project managers from key projects as and when required

- Senior managers from key suppliers (if appropriate at this stage)

4.3.8 Stakeholder/communications manager

The stakeholder/communications manager's role includes establishing an understanding of the programme's stakeholders and a plan determining either their level of engagement or how they will be kept informed of the progress of the programme. This will involve the following:

- Identifying all stakeholders

- Carrying out an influence/impact analysis

- Developing a communication plan for stakeholders

- Developing a communication plan for staff within the client organisation affected by the programme

- Developing a public relations plan for the programme

4.3.9 Programme management office

In addition to requiring a programme manager, this stage also requires that the core members of the programme management office be available. These include the head of the programme management office and sufficient staff with the necessary technical expertise to carry out the functions required during this stage to develop the programme delivery plan and establish the governance and planning controls.

The size and range of expertise required will depend on the nature and complexity of the programme. In some instances, the input required from the programme team may be available from within the client/sponsoring organisation, but in many others, it is likely that all the expertise will be externally sourced specifically for the programme (see Figure 4.20).

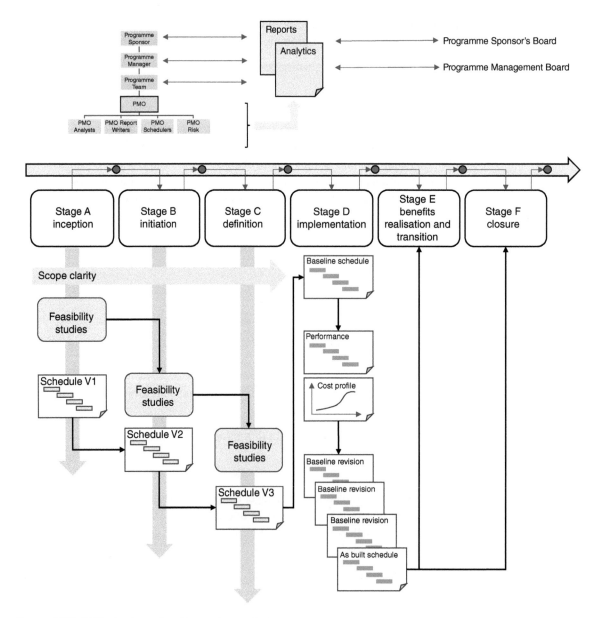

Figure 4.20 PMO structure, function and reporting.

In addition to specialist input on matters such as legal aspects, property and finance, some sponsoring organisations may need to contract with organisations to provide all or part of the programme management delivery capability. These positions include the following:

- Head of programme management office

  - Responsible for setting up the programme management office and ensuring all systems, processes and procedures are established in readiness for the implementation of the programme.

- Scheduling manager

  - Responsible for establishing the planning and schedule infrastructure to be adopted across the programme and its projects and for developing the overall programme schedule.

- Cost manager

  - Responsible for establishing the cost management system and protocols to be adopted across the programme and its projects and for developing the cost budget and cash flow forecasts.

- Risk manager

  - Responsible for establishing the risk management system to be adopted across the programme and for carrying out the process to determine a risk analysis for the programme.

- Document/information/IT security manager

  - Responsible for determining, in conjunction with the programme manager, the document and information system to be rolled out across the programme and for installing this system in preparation for implementation, including ensuring a secure information storage and exchange mechanism.

- Data manager

  - Responsible for data collection, data system management, data reporting and analysis and ensuring data collaboration among all the participants in data exchange.

- Administrator(s)

  - Responsible for providing sufficient support to allow the efficient and effective operation of the programme management office.

- Health, safety and quality manager

  - Responsible for setting up a system that is appropriate for the health, safety and quality regulations and legalisation existing in the context in which the programme is being carried out.

- Sustainability and environmental manager

  - Responsible for establishing from the programme brief the sustainability drivers, identifying any statutory and regulatory requirements affecting the programme and developing a framework for sustainability targets.

- Specialist advisors

  - During this stage, it is likely that the project management team will need to obtain the advice and assistance of specialist advisors. These areas could be legal, real estate, financial or a range of technical aspects.

5 Stage D: Implementation

5.1 Purpose of this stage

The purpose of this stage is the initiation and execution of the various projects comprising the programme, including assessing the performance of individual projects, managing the interfaces between projects, monitoring benefits realisation, managing financial expenditure and managing the introduction of any changes to the programme.

Please refer to the Programme Delivery Matrix (PDM) in Appendix A to understand the key roles and artefacts that comprise the CIOB programme life cycle.

5.2 Key activities of this stage

In this stage, the physical activities of the programme are executed through a number of projects. The programme progress report is a key artefact that describes the level of progress being achieved by the programme team (see Figure 5.1).

Key artefacts	CIOB code of practice programme management stages
Vision statement	Stage A - Inception
Programme mandate	Stage A - Inception
Programme brief	Stage B - Initiation
Programme business case	Stage B - Initiation
Programme delivery plan	Stage C - Definition
Programme progress report	Stage D - Implementation
Programme benefits report	Stage E - Benefits realisation and transition
Programme closure report	Stage F - Closure

Figure 5.1 Key artefacts: Stage D (implementation).

The focus of the programme management team is on ensuring the successful implementation of each individual project in accordance with the planned sequence and the delivery of the required outputs (see Figure 5.4).

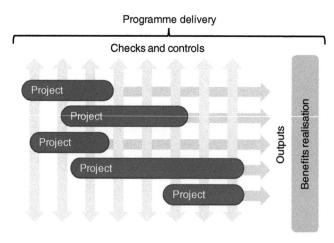

Figure 5.2 Stage D: implementation.

Procedural and management activities as defined during the earlier stages will be put into use to monitor and control the implementation of the outputs and outcomes (see Figure 5.2).

It is the responsibility of the programme manager to ensure that resources across the programme are managed effectively. This will include the utilisation of internal and external resources and resources that may be shared across projects and programmes. Furthermore, the enabling technology and tools identified during earlier stages to support the implementation should also be made available. It is often advisable, particularly for large and complex programmes, to have core teams co-located or at least undergo team-building exercises to develop effective relationships and engender trust.

In addition to the core programme management activities identified later, the following activities, which are initiated during the early stages of the programme, need ongoing attention during the implementation phase to ensure desired delivery of the programme outcomes and benefits:

- Programme delivery plan

- Benefits management

- Stakeholder management

- Business case management

- Transition management

5.2.1 Initiate projects

- Appoint the initial project management teams

- Issue terms of reference and brief the project teams

- Liaise with the project management office with the governing structures to establish processes and procedures

5.2.2 Performance monitoring and control

Implement the management processes and procedures to monitor and control the programme in the areas of:

- Stakeholder management

- Communications

- Financial management

- Benefits realisation

- Governance

- Risks/issues

- Time

- Costs

- Performance

- Quality management

- Monitor the progress of implementation against the programme delivery plan (PDP)

5.2.3 Reporting

- Report regularly to the programme sponsor/programme sponsor's board on the progress of programme delivery

- Raise any critical issues requiring consideration by the programme sponsor's board

5.2.4 Project closure

- Manage the formal closure of projects, validate project outputs and take ownership of outputs/outcomes

- Carry out end-of-project reviews and compare actual achievement with the required outcomes and benefit realisation

- Carry out a lessons-learned review and highlight any aspects that will benefit other projects

5.2.5 Construction procurement methods

In the built environment, there are numerous methods of procuring construction work. It is important that the various options are studied so as to ensure that the right strategy is adopted that is in line with the programme outcomes (see Figure 5.3).

5.2.6 Measuring progress

It is important that the programme team understands the criticality of the projects being managed and how they could impact the overall completion date (see Figure 5.4).

All Gantt charts can be converted into S curves. These curves provide an indication of the cumulative level of effort that will be required to deliver. They also provide a one-page summary for effective management reporting.

5.2.7 Data collection and analysis

The programme management office should determine in advance the level and type of detail required. Data should not be collected for the sake of collecting data.

Data is passed up the programme hierarchy to the programme management office. Occasionally, the programme management office must perform some independent analysis to see if the right performance is being achieved by the entire programme team (see Figure 5.5).

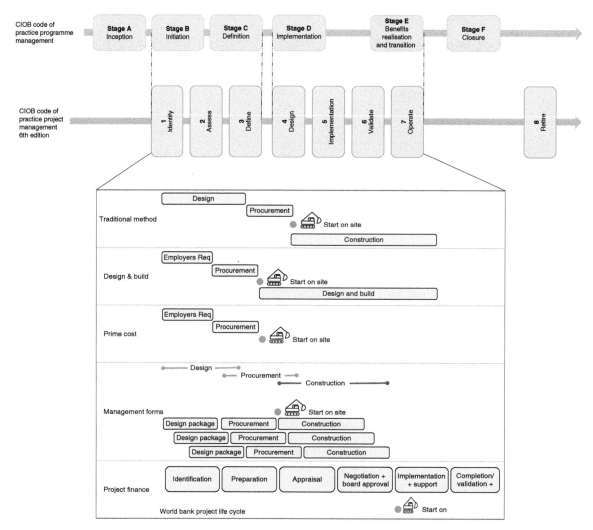

Figure 5.3 Common construction procurement methods.

5.2.8 Data management

It is important that a sensible approach be designed and developed for the programme. Collecting data for the sake of it is not efficient and creates unnecessary administrative burdens for the programme team (see Figure 5.6).

The optimal amount of data must be collected from the programme teams so that meaningful insights can be deduced from performance being achieved.

Constant refinement of the data collection process should be conducted by the programme management office to ensure that data is meaningful and insightful.

5.2.9 Monitoring progress

It is not possible to measure everything that happens on a programme. Even if elaborate systems are in place that collect and aggregate data from the projects, it is imperative that the programme management office have the ability to see what the overall status is.

The programme team and stakeholders must be able to quickly understand the performance being achieved and what further action is required.

In Figure 5.7, the same information can be enhanced by showing acceptable ranges that provide more meaning to what is being described.

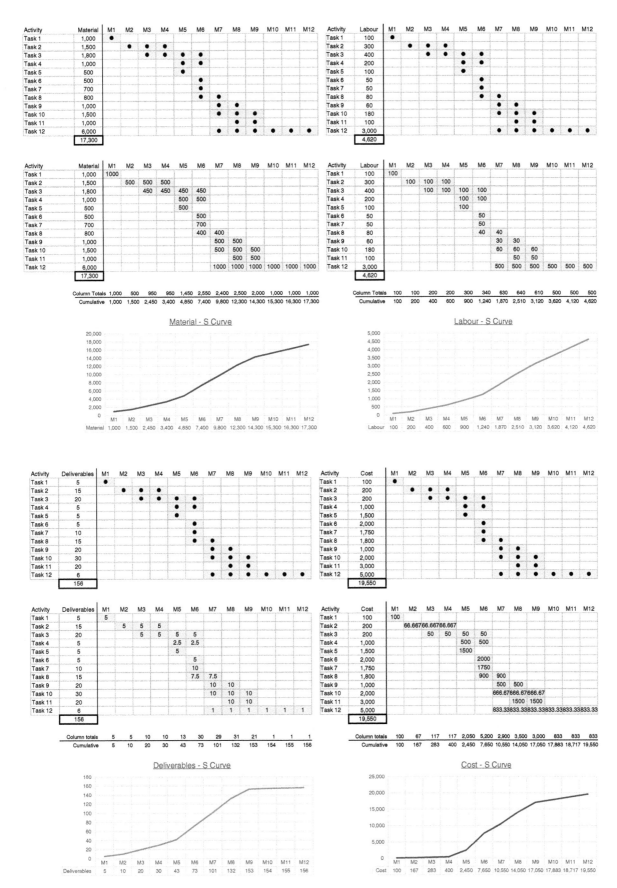

Figure 5.4 Measuring progress using S curves.

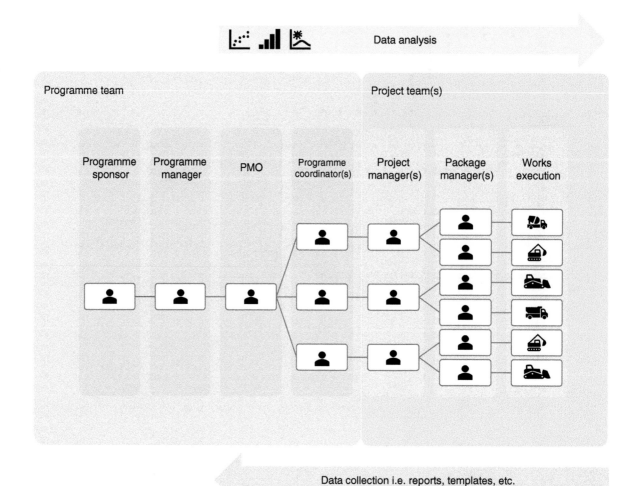

Figure 5.5 Data analysis on programmes.

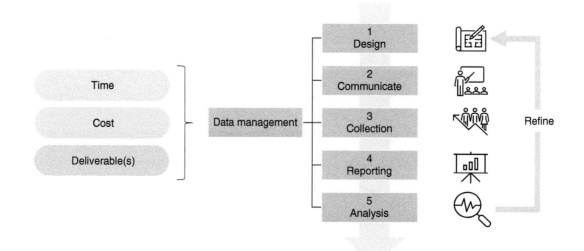

Figure 5.6 Data collection on programmes.

	Forecast	Cumulative forecast	Actual	Cumulative actual	Lower limit			Upper limit		
					100%	85%	70%	110%	130%	150%
					Green zone	Amber zone	Red zone	Green zone	Amber zone	Red zone
Jan	50	50	20	20	50	43	35	55	65	75
Feb	40	90	10	30	90	77	63	99	117	135
Mar	50	140	10	40	140	119	98	154	182	210
Apr	65	205	50	90	205	174	143.5	225.5	266.5	307.5
May	60	265	50	140	265	225	185.5	291.5	344.5	397.5
Jun	130	395			395	336	276.5	434.5	513.5	592.5
Jul	200	595			595	506	416.5	654.5	773.5	892.5
Aug	350	945			945	803	661.5	1039.5	1228.5	1417.5
Sep	400	1345			1345	1143	941.5	1479.5	1748.5	2017.5
Oct	500	1845			1845	1568	1291.5	2029.5	2398.5	2767.5
Nov	350	2195			2195	1866	1536.5	2414.5	2853.5	3292.5
Dec	300	2495			2495	2121	1746.5	2744.5	3243.5	3742.5

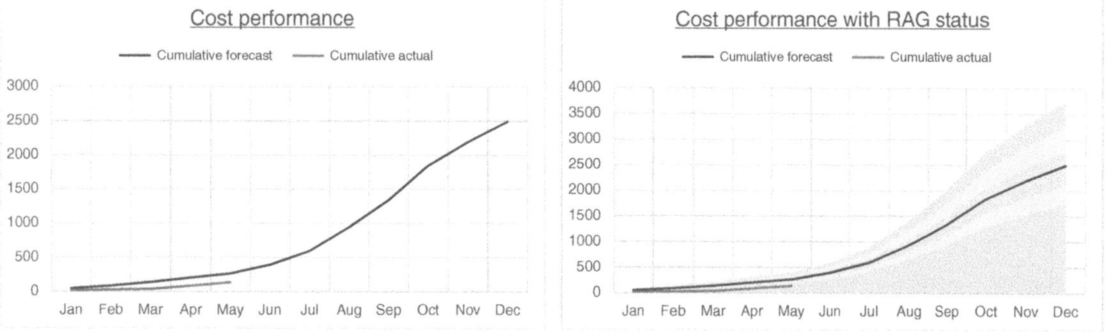

Figure 5.7 Measuring progress on programmes.

5.2.10 Managing multiple projects

At the programme level, the performance of projects must be shown in a clear and concise manner. An example of how a group of projects can be reported on is shown in Figure 5.8.

This type of reporting allows the programme team to focus on those projects that require attention rather than reviewing each and every one.

5.2.11 Reporting and forecasting

Programme schedules should contain a range of forecasts so that the criticality can be conveyed to the programme team and stakeholders.

Figure 5.9 shows a typical project. The contractual duration and associated extensions of time are shown on the schedule. In addition to this, different forecasts are shown from the contractor through the programme manager. This provides a range of forecasts and provides a better insight into where the project could finish.

5.2.12 Programme delivery plan

Managing the PDP during delivery is vital to ensure that, as the programme progresses, projects remain integrated with each other and the programme. In practice, this means regularly reviewing the PDP with the programme manager, programme business change manager or their representatives and the project managers. It is important to give project managers the opportunity to report progress and programme business change managers the opportunity to report operational issues. One way of doing this might be by undertaking PDP review workshops.

Holding a workshop with all project managers and change managers present will ensure that the knock-on effects of changes or ideas in one project or business area are explored with respect to the PDP in its entirety.

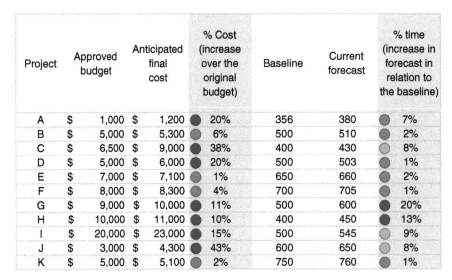

Project	Approved budget	Anticipated final cost	% Cost (increase over the original budget)		Baseline	Current forecast	% time (increase in forecast in relation to the baseline)	
A	$ 1,000	$ 1,200	●	20%	356	380	●	7%
B	$ 5,000	$ 5,300	●	6%	500	510	●	2%
C	$ 6,500	$ 9,000	●	38%	400	430	○	8%
D	$ 5,000	$ 6,000	●	20%	500	503	●	1%
E	$ 7,000	$ 7,100	○	1%	650	660	●	2%
F	$ 8,000	$ 8,300	●	4%	700	705	●	1%
G	$ 9,000	$ 10,000	●	11%	500	600	●	20%
H	$ 10,000	$ 11,000	●	10%	400	450	●	13%
I	$ 20,000	$ 23,000	●	15%	500	545	○	9%
J	$ 3,000	$ 4,300	●	43%	600	650	○	8%
K	$ 5,000	$ 5,100	●	2%	750	760	●	1%

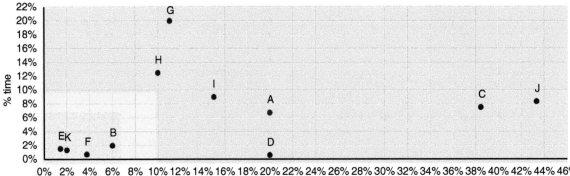

Figure 5.8 Reporting on multiple projects in a programme.

Activity ID	Activity Name	Start	Finish	At Completion Duration
Project A		01-Oct-22 A	31-Mar-24	548
Contractual		01-Oct-22 A	31-Dec-23	456
A001	Construction duration	01-Oct-22 A	31-Dec-23	456
A002	Contract duration	01-Oct-22 A	05-Sep-23	340
A003	EOT 1	06-Sep-23	04-Oct-23	28
A004	EOT 2	04-Oct-23	29-Nov-23	56
A005	EOT 3	29-Nov-23	31-Dec-23	32
Forecasts		01-Oct-22 A	31-Mar-24	548
A1000	Contracor's forecast	01-Oct-22 A	31-Jan-24	488
A1010	Project Manager forecast	01-Oct-22 A	29-Feb-24	517
A1020	PMO forecast	01-Oct-22 A	29-Feb-24	517
A1030	PMO forecast - RCF	01-Oct-22 A	31-Mar-24	548
A1040	PMO foecast - Risk Adjusted	01-Oct-22 A	31-Mar-24	548
A1050	Programme Manager forecast	01-Oct-22 A	31-Mar-24	548

Forecast summary

Contractual	456
Contractor's forecast	488
PM forecast	517
PMO forecast	527
PMO RCF	548
PMO Risk Adjusted forecast	548
Prg Man forecast	548
Average	518.86

Contractual	456
Contractor's forecast	488
PM forecast	517
PMO forecast	527
PMO RCF	548
PMO Risk Adjusted forecast	548
Prg Man forecast	548
Average	518.86

Figure 5.9 Forecasting schedule information.

The regularity of PDP workshops will depend on what has been agreed on during the definition phase, but a minimum should be undertaken as each step change in the programme is implemented.

The results of the review will usually mean that the programme delivery has to be updated and signed off before the next step change can be planned.

5.2.13 Benefits management

Benefits need to be regularly reviewed during the implementation stage as:

- Unforeseen issues and risks arise that require a change of plan

- Strategic priorities change in response to external circumstances

- Assumptions, and/or constraints, that were used to determine potential benefits have been discovered to be inaccurate

The method of review has been determined during the definition stage.

The benefit profiles will need to be updated, and the benefit realisation activities in the PDP will need to be amended to reflect any changes.

5.2.14 Business case management

The business case is the basis upon which the implementation phase of the programme is commissioned. As delivery progresses, changes outside the control of the programme may mean that the assumptions and estimates upon which the business case was made change.

The business case should be reviewed periodically to reflect the performance of the projects and programme. It is the responsibility of the programme sponsor during implementation to ensure that the business case is updated and, if the case is fundamentally affected, to explore changes to the PDP and benefits in order to maintain the case.

If projects are not viable, the programme might be re-assessed.

Conversely, it is also possible that changed circumstances may lead to further strengthening of the business case. In practical terms, this will mean looking for opportunities to increase the benefits and reduce the costs. In some cases, increasing the cost may facilitate greater benefits. However, this should be balanced against making too many changes that lead to unproductive use of resources or increased risk of not achieving the desired outcomes.

5.2.15 Performance monitoring, control and reporting

The amount and level of detail required for progress reporting vary widely from organisation to organisation and programme to programme. The approach to be taken, including the documentation to be used, was defined during the definition phase (see Figure 5.10).

It is important that the information provided is complete, timely and accurate so that appropriate decisions can be made by the programme sponsor's board. These decisions include the following:

- Decisions taken with respect to changes in the macro environment

- Review of outcomes, including any further actions necessary

- Review of realisation of benefits, including any further actions necessary

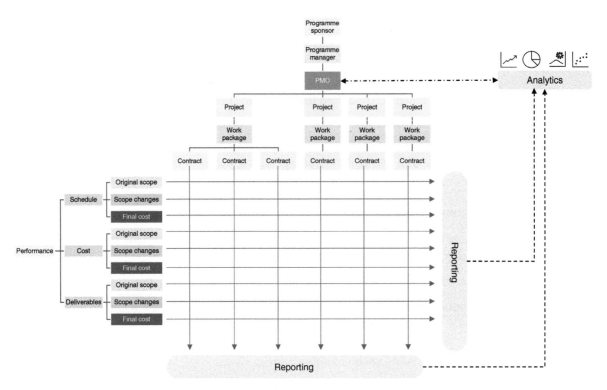

Figure 5.10 Programme monitoring and control.

- Review of actual progress against planned progress and dealing with variance
- Review of the financial position for the programme and dealing with variance

Typically, the programme manager will produce a programme highlight report, which will form the basis of the review. An indicative template of this report is shown in Appendix C.

The programme team (see Figure 5.11) must ensure that the information being received from projects is:

- Relevant
- Timely
- Accurately depicts the real performance being achieved
- Is not too bureaucratic in nature and can be produced quickly

It is imperative that the programme team independently assess the performance of projects within the programme. It is customary to measure performance against what was originally contracted for. Therefore, there is an emphasis on actual versus the planned.

However, some independent analysis should be undertaken from time to time to understand what the underlying causes of below-average performance are. Basic statistics, correlation and regression techniques can be used to understand performance on a programme (see Figure 5.12).

5.2.16 Correct use of graphics

Correct use of graphs is important so that the meaning can be conveyed to the end user in a way that enhances understanding and promotes decision-making.

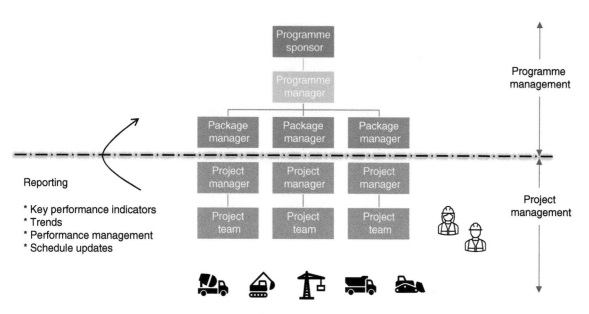

Figure 5.11 Interface between programme and project reporting.

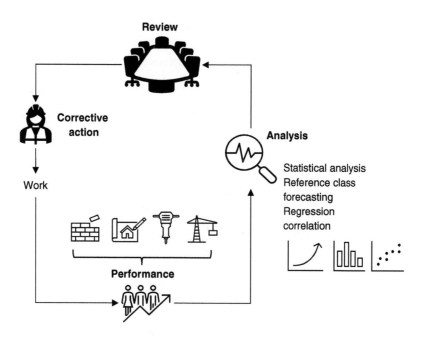

Figure 5.12 The need for independent analysis.

	Forecast	Actual	Forecast - cumulative	Actual cumulative	Forecast to complete
Jan	30	10	30	10	
Feb	20	10	50	20	
Mar	30	20	80	40	
Apr	40	30	120	70	
May	50	30	170	100	100
Jun	60		230		150
Jul	30		260		200
Aug	35		295		250
Sep	40		335		300
Oct	50		385		350
Nov	60		445		400
Dec	20		465		465

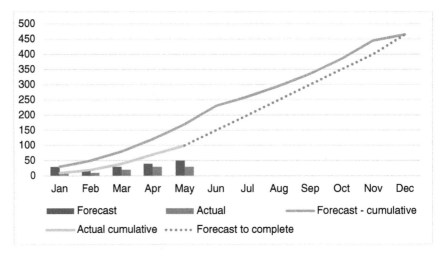

Figure 5.13 Correct use of graphics (example 1).

For example, sometimes too much data on the same graph can conceal what is really going on. Figure 5.13 shows a typical example.

In the above graph, what can be seen is that the programme is not achieving its forecasted targets. However, the forecast to be completed shows that it will still be completed according to the original target.

However, we can look at the underlying data differently to see how progress is being achieved (see Figure 5.14). This clearly shows that the forecast to complete is outside the original target.

5.2.17 Risk and issue management

Risk and issue management are ongoing programme management functions. The programme's approach to risk and issue management will follow the process defined during the definition phase.

Risk and issue registers should be regularly maintained and monitored. The actions agreed upon either to mitigate the risk or resolve the issue should be included in the overall PDP or monitored alongside the PDP.

It is important to ensure that, where necessary, risks that look as if they are going to materialise and issues that cannot be resolved are escalated. Who to escalate to and when will be documented in the approach defined during the definition phase.

		A	B	C	D	E	F	G	H
		Forecast	Actual	Actual %	Forecast - cumulative	Actual cumulative	Future performance	Predicted actuals	Forecast to complete
2022	Jan	30	10	33%	30	10			
	Feb	20	10	50%	50	20			
	Mar	30	20	67%	80	40			
	Apr	40	30	75%	120	70			
	May	50	30	60%	170	100			100
	Jun	60			230		57%	34	134
	Jul	30			260		57%	17	151
	Aug	35			295		57%	20	171
	Sep	40			335		57%	23	194
	Oct	50			385		57%	29	223
	Nov	60			445		57%	34	257
	Dec	20			465		57%	11	268
2023	Jan							20	288
	Feb							20	308
	Mar							17	325

Min	33%
Max	75%
Mean	57%

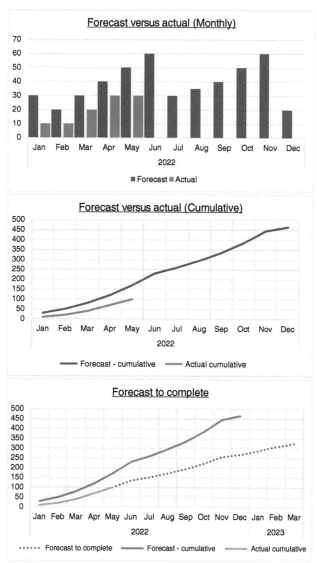

Figure 5.14 Correct use of graphics (example 2).

Risks and issues are often recorded in one or more closely related documents. Whichever method is used, it is important to recognise that managing risks and issues are two separate processes. Issues may arise that need to be resolved before formal risk and issue review meetings. It is important that arrangements are put in place to immediately deal with issues as they arise. It is also possible that some issues may have an influence on the fundamental business case assumptions, in which case the programme manager should ensure that a considered resolution of the issue is undertaken.

5.2.18 Financial management

The financial status of the programme should be monitored in accordance with the process defined during the definition stage. This monitoring will include:

- Ensuring that appropriate financial controls are in place for the programme

- Ensuring that the projects and overall programme are being managed within budget

- Providing timely reports to the programme sponsor's board on financial status and highlighting any significant variance from the budget or changes that might incur unplanned additional expenditure (this can be included in the programme highlight report – see Appendix C)

- Ensuring that the financial processes in place for the programme comply with corporate standards and regulations

- Maintaining up-to-date documentation on the financial status of the programme and the projects that are helping to deliver the overall programme

5.2.19 Change management

Change control is the formal identification, evaluation and decision-making required for accepting or rejecting changes to aspects of a programme. The management of change during the implementation phase should follow the approach defined during the definition phase.

The principle is usually that if a change can be managed at a project level, then this should be the preferred way of managing a change.

Changes to the PDP, business case, benefits and budget will affect the overall programme and should be evaluated and approved at the level of the programme sponsor's board.

The following types of information and signed-off documents should typically be subject to change control:

- Core programme documents

- Programme governance

- Information on data protection issues

- Commercial and human relations-sensitive information

- Programme communications

- Project-related information

A change can have a positive as well as a negative influence on the business case, and unless managed appropriately, it may ultimately lead to detrimental effects towards the realisation of the benefits envisaged.

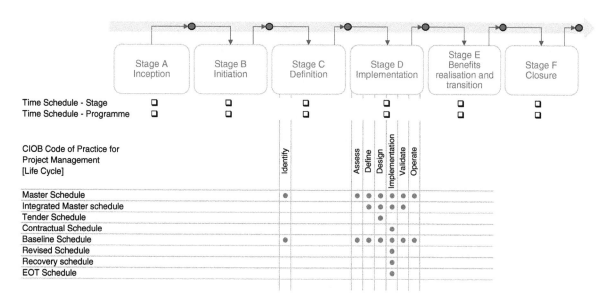

Figure 5.15 Programme time management.

5.2.20 Time management

A sensible and practical structure must be agreed upon to manage time. In the built environment, we have numerous types of schedules, and it is important that, at the outset, a clear classification is developed, maintained and communicated to all parties.

Figure 5.15 shows the different types of time schedules and where they might be encountered.

5.2.21 Information management

Appropriate information management arrangements should be put in place in accordance with the agreed-upon approach defined during the definition phase. This will ensure that information is available when it needs to be, is current, and is compliant with the quality standards set for the programme.

5.2.22 Stakeholder/communications management

During the implementation stage, it is essential to continue with planned engagement and communication activities with stakeholders as well as respond to unforeseen events.

The stakeholder engagement and communications plan should be regularly reviewed and adjusted to take into account any changes to planned delivery that need to be communicated.

The programme business change manager and programme manager should satisfy themselves that people with responsibility for engaging and communicating with stakeholders are doing so. Updates on progress can be included in the highlight reports.

Stakeholder engagement and communications are also key parts of the programme leadership function. The programme sponsor's board and programme sponsor should champion the programme in day-to-day communication. They should also be sensitive to potential issues at a strategic level and respond accordingly where necessary with input from the programme manager.

5.2.23 Quality management

The implementation stage of the programme ensures that the agreed method for assessing quality assurance and fitness for purpose, as agreed during the definition phase, is put into practice. This may include activities ensuring that for each project output or programme deliverable, acceptance or performance criteria are in place and are assessed, monitored and reported.

If part of the agreed quality or assurance checks is to undertake periodic reviews at key points in the programme's implementation, these will need to be planned, organised and undertaken. Any findings and recommended actions would be fed into the overall PDP to improve performance.

The programme manager must ensure that when deciding quality and acceptance criteria for deliverables and project outputs, consideration is given to the practicalities of assessing them so that the assessment data can be tested for 'fitness for purpose'.

5.2.24 Procurement and commercial management

The procurement and commercial management of programmes (and their components) will depend on the nature and type of the programme, its funding arrangements and whether it is in the public sector or in private sector. In some instances, the procurement process may also include obtaining the services of an external programme manager and programme management office.

Regardless of the type and approach, the procurement and commercial management processes at the implementation phase would need to consider the following:

- Contract and relevant documentation are in place
- Contractual arrangements are up-to-date
- Business case is valid and up-to-date
- Original projected business benefit remains feasible
- Processes and procedures are in place to ensure the achievement of outputs, outcomes and benefits
- PDP contains processes embedded to ensure all necessary acceptances and fit-for-purpose testing, such as commissioning and soft landings
- Business contingency, continuity and/or reversion arrangements are in place
- Risks and issues are being managed
- Adequacy and availability of resources
- The delivery plans are still feasible
- There are management and organisational controls to manage the project through delivery and use
- Arrangements are confirmed for handover of the project from the programme sponsor to the customers/clients/end users
- There are agreed-upon plans for training, communication, rollout, production release and support as required
- Governance plans are in place
- Information management plans are in place

- Those involved in procurement and commercial management functions need to produce SMART (Specific, Measurable, Achievable, Realistic and Time bound) outputs or outcomes and have a clear understanding of the contract

- The specification, performance measures and contractual terms should be well defined and clear to all relevant stakeholders

- Effective teamwork and relationship management need to be in place, and this is especially important in complex programmes due to their long-term nature

5.2.25 Health and safety management

At the implementation stage, the overall health and safety management plan, as prepared during the earlier stages, will be put into action. Typically, all the projects and activities undertaken during the implementation phase will adhere to the health and safety management plan requirements, particularly the targets and key performance indicators that may have been set at the programme level. The highlight report for the programme may contain a summary of the health and safety performance of the projects and activities, reporting on the actual progress against the targets and key performance indicators.

Project-level health and safety management will be a key element in ensuring the achievement of the overall programme targets.

5.2.26 Sustainability/environmental management

The overall programme sustainability and environmental management parameters and targets will be set at the earlier stages. The processes and procedures, including overall programme targets and key performance indicators, must be incorporated into individual projects and activities. Projects that may require planning permission will prepare environmental impact assessments that will also encompass the programme's sustainability and environmental management aspirations.

5.2.27 Transition management – projects closure

Transition plans are necessary to successfully translate deliverables and outputs into outcomes that will enable the realisation of benefits envisaged within the programme business case. These can be included as an update to the overall PDP. Transition plans set out the tasks needed to ensure that the new capability will be embedded into business as usual operations. They might include:

- Training for the new capability

- Providing guidance on the new process

- Allowing parallel running of old and new processes

- Setting up temporary support

- Soft-landing the new capability

- Setting up monitoring processes so that there is assurance that the new ways of working are being used

- Benefit summary and tracking post programme closure (i.e. some benefits are validated post closure, for example, user's satisfaction)

It is important to note that inadequate transition management may hamper the successful outcome of the programme even if the project outputs are delivered successfully.

5.3 Key roles and responsibilities of this stage

The implementation stage marks the maximum effort and involvement of the full programme management team, as there is a collective effort to deliver the outcomes of the programme in accordance with the proposals set out in the PDP. This effort is headed by the programme manager, who, closely supported by the programme management office, initiates, monitors and closes the projects, which make up the programme. The business case manager, together with the benefits realisation manager, ensures that the required benefits are being progressively realised and that the new capabilities or facilities are being effectively introduced into the client's organisation.

5.3.1 Programme sponsor's board

The programme sponsor's board continues its role as adviser and approver, fulfilling the following duties:

- Reviewing regular reports on the programme's progress

- Providing overview of the progressive realisation of benefits

- Resolving any issues raised by the programme sponsor

- Providing any approvals or decisions on matters raised by the programme sponsor

- Providing overview of the introduction of any client-generated changes

5.3.2 Programme sponsor

Continuing to be the interface between the programme and the client's organisation, the programme sponsor maintains a check on the programme's overall progress and keeps the client advised of this progress. Principal tasks include:

- Regularly reviewing progress towards the programme's objectives

- Regularly monitoring the performance of the project management team

- Referring any major issues to the programme sponsor's board

- Obtaining any approvals/decisions required from the programme sponsor's board

- Verifying the programme manager's recommendation for appointment of project managers

- Dealing with any client requirements for changes to programme scope, objectives or deliverables

- Determining, in conjunction with the programme manager, any additional projects required to maintain (or enhance) the planned deliverables

- Depending on the nature and complexity of the programme, it may be necessary for the programme sponsor and/or programme manager to initiate either internal or external audits of the programme or projects

5.3.3 Programme business change manager

The programme business change manager is focused on ensuring that the programme is achieving its planned objectives by:

- Monitoring achievement of deliverables against the plan

- Managing the integration of the changes into the client's organisation

- Keeping aware of any changing external circumstances that may affect the outcomes of the programme and impact benefits

5.3.4 Programme benefits realisation manager

Concerned with ensuring that the programme is delivering benefits in accordance with the PDP, the benefits realisation manager has the following responsibilities:

- Monitoring benefits being delivered by projects against the criteria set out in the benefit profiles

- Reviewing benefit realisation against the key success criteria

- Assessing the effectiveness of the benefits being realised

- Reporting any variances in benefits being realised to the programme business change manager/programme manager

5.3.5 Programme manager

With full responsibility for the implementation of the programme, the programme manager has a wide range of tasks, including the following:

- Initiating projects in accordance with the PDP

- Selecting and appointing project managers

- Approving the selection of project teams and consultants

- Monitoring the progress on individual projects

- Monitoring the performance of project teams

- Monitoring the quality of outputs from projects

- Approving the closure of projects

- Liaising with the programme business change manager on transition arrangements at the closure of each project

- Resolving any issues on projects

- Raising any major issues with the programme management board

- Reviewing regularly progress of the programme in terms of time, cost, risks, deliverables, etc. with the programme management office

- Reviewing regularly stakeholder issues with the stakeholder/communications manager

- Reviewing regularly financial issues with the programme financial manager

- Reviewing regularly benefits realisation with the programme business change manager/programme benefits realisation manager

- Reviewing regularly health and safety issues with the health and safety manager

- Reviewing regularly sustainability issues with the sustainability manager

- Reporting regularly the progress of the programme to the programme sponsor

- Referring any critical issues to the programme sponsor

- Managing the introduction of any changes instructed by the programme sponsor

5.3.6 Programme financial manager

Throughout the implementation stage, the programme financial manager continues to manage the financial aspects of the programme. The tasks of the programme financial manager will include:

- In conjunction with the programme manager, monitors and reports to programme sponsor and client on programme expenditure and the predicted final cost

- Liaises with funder(s) on the release of monies

- Manages risk and contingency allowances

- Reviews the programme's financial performance against the business case

- Monitors the cost implications of instructed client changes

- Reports any significant issues/problems to the programme manager

5.3.7 Programme management board

The programme management board continues to provide advice and support to the programme manager during programme implementation. The programme management board:

- Reviews the programme's progress regularly

- Resolves any issues that are impacting progress

- Refers any significant issues to the programme sponsor's board for advice or resolution

5.3.8 Stakeholder/communications manager

The stakeholder/communications manager continues to ensure interested stakeholders are identified and that there is a mechanism for maintaining their engagement in the objectives of the programme by doing the following:

- Reviewing regularly the stakeholder analysis

- Maintaining regular communication with stakeholders

- Maintaining regular communication with the client organisation

- Maintaining the activities identified in the public relations plan

- Reporting any significant issues/problems to the programme manager

5.3.9 Programme management office

The programme management office is responsible for providing the technical functions that allow the programme to be managed and controlled, including the following tasks:

- Ongoing monitoring of progress on programme and projects

- Ensuring projects are being managed in accordance with the processes and procedures defined in the PDP

- Preparing regular reports on progress for programme management board and the programme sponsor

- Maintaining change control process

- Maintaining issues log and highlighting major issues to programme manager

Within the programme management office, functions with specific responsibilities include:

- Head of programme management office
  - Maintains the ongoing operation of the programme management office
- Scheduling manager
  - Monitors progress on the programme and all projects
- Cost manager
  - Maintains cost monitoring and control on the programme and projects
- Risk manager
  - Maintains the risk management process
- Document/information manager
  - Maintains the programme's document and information system
- Health and safety manager
  - Ensures the programme and projects comply with all relevant health and safety regulations and legalisation
- Sustainability manager
  - Ensures the programme and projects comply with all relevant sustainability regulations and legalisation and monitors progress against sustainability targets
- Administrator(s)
  - Provides support to the programme management office

5.3.10 Project management structures

Each project that is carried out as part of the programme will need its own management and technical teams responsible for the execution of the project. These teams are likely to include the following:

- Project managers
- Project support office
  - Head of programme support office
  - Scheduling manager
  - Cost manager
  - Risk manager
  - Document/information manager
  - Administrators
- Consultants and contractors
  - Design team
  - Contractor
  - Suppliers
  - Health and safety

5.3.11 Key points of contact

To ensure smooth communication and minimise any misunderstandings, it is important that the key points of contact are maintained throughout the life of the programme (see Figure 5.16).

The programme sponsor manages upwards through the client organisation to ensure that all the relevant approvals are obtained.

The programme manager manages the programme team.

Team members from within the programme team may be tasked with being the point of contact for the project managers delivering the individual projects.

It is important that a formal line of communication exists so that miscommunication and confusion is avoided.

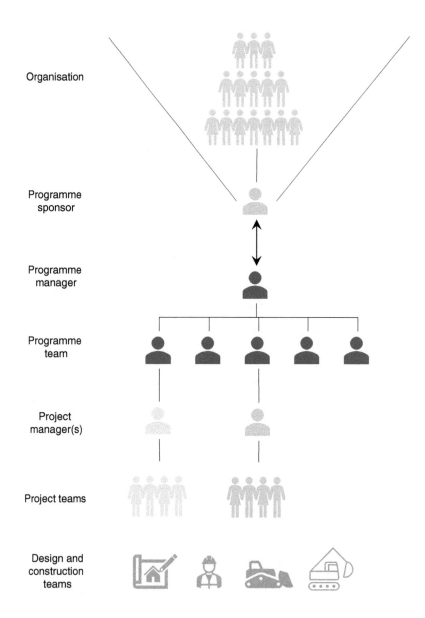

Figure 5.16 Key points of contact.

6 Stage E: Benefits Realisation and Transition

6.1 Purpose of this stage

At the point at which the deliverables of a programme are considered by the programme manager and programme sponsor to be achieved, a review is undertaken by the programme business change manager to assess if the required outputs and final outcome have been realised. This is a key distinction between the linear nature of a project and what can be a more iterative approach in a programme between stages C (Definition), D (Implementation) and E (Benefits Realisation and Transition).

This review also considers the success of the transition of programme outputs into the final permanent state of the new undertaking. This review will confirm whether a programme can be shut down or if further activity needs to be carried out. It is possible that any further activity can be executed without the need to retain the programme management structure (see Figure 6.1).

Please refer to the Programme Delivery Matrix (PDM) in Appendix A to understand the key roles and artefacts that comprise the CIOB programme life cycle.

6.2 Key activities of this stage

The programme benefits report is a key document that provides an update on progress relating to benefits during this stage (see Figure 6.2).

6.2.1 Benefits review

The programme sponsor, programme manager, programme business change manager and the programme business realisation manager review the outcomes to ensure that what the programme has delivered is what was stipulated in the programme brief, business case and programme delivery plan.

An assessment is made by the programme sponsor and programme business change manager as to whether the outcomes delivered by the programme will achieve the benefits required. Their assessment is referred to the programme sponsor for ratification.

There is now an opportunity to make any final adjustments to what has been delivered in the light of greater knowledge and any changing circumstances.

If the programme sponsor's board confirms the programme has delivered what is required, then they instruct the closure of the programme.

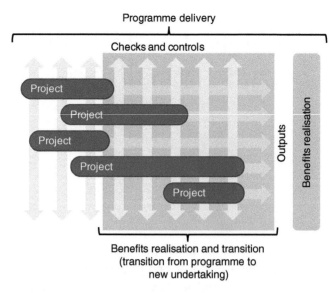

Figure 6.1 Stage E: Benefits review and transition.

Key artefacts	CIOB code of practice programme management stages
Vision statement	Stage A - Inception
Programme mandate	
Programme brief	Stage B - Initiation
Programme business case	
Programme delivery plan	Stage C - Definition
Programme progress report	Stage D - Implementation
Programme benefits report	Stage E - Benefits realisation and transition
Programme closure report	Stage F - Closure

Figure 6.2 Key artefacts: Stage E (Benefits realisation and transition).

6.2.2 Benefits realisation

Realisation of benefits may occur over an extended period as projects become completed, and in some instances, the full realisation of the benefits arising from the programme may take months or years beyond the completion of the programme. The situation may therefore arise where it is possible to consider partial completion of a programme and to disband part of the programme team. In these circumstances, the programme business change manager will need to maintain an ongoing measurement of the benefits. Also, in some instances, this may require the programme sponsor's board to be in place until full and final realisation is achieved.

6.2.3 Transition

Transition from the programme to the new enterprise can be progressive, as elements of the new undertaking are handed over as projects complete or be a total handover that occurs at the completion of the programme. The new permanent operational state of the undertaking requires a management structure and resources to be mobilised ready to receive and commission it. The transition arrangements are overseen by the programme business change manager.

6.2.4 Training and induction

When the outputs from projects are physical facilities, arrangements must be made for the staff of the new enterprise to receive induction training in the running and maintenance of the new facility. In support of maintaining the new facilities, there should be a transfer of project information, such as design details, building information, modelling information, operation and maintenance manuals, legal agreements and design warranties.

6.2.5 Benefits management

Benefits management is a critical activity in any programme, regardless of its type, objectives and duration. There have been many programmes that delivered great outputs and capabilities but failed to realise benefits due to insufficient or inappropriate arrangements being made to ensure that benefits were realised.

6.2.6 Managing and realising benefits

Benefits management and realisation are core elements of programme (and change) management. It provides a systematic approach to identifying, defining, tracking, realising, optimising, reviewing and communicating benefits, during and beyond a programme.

It is essential that, from the onset, the programme sponsor takes ownership of the benefits agenda and throughout the life of the programme provides strong leadership, particularly in terms of prioritising benefits, with support from the programme business change manager. Many programmes give in to the temptation of producing a long list of anticipated benefits in the business case in order to maximise the chances of securing funding. Once the funding is achieved, the willingness and commitment to take ownership of all the promised benefits diminishes – the programme sponsor and programme business change manager should be aware of this and must examine and prioritise the benefits, including assigning individual ownership and seeking commitment from individual benefit owners where appropriate, so that benefits realisation is monitored and delivered as anticipated.

The objectives of putting in place processes to manage and realise benefits in a structured way include the following:

- Identification of benefits

- Defining the benefits

- Alignment of the benefits to the strategic objectives, programme vision and outcomes

- Assigning appropriate ownership of benefits, including their realisation

- Aligning project outputs to support the benefits realisation

- Identifying and undertaking business changes that will be necessary to deliver the benefits

- Monitoring and evaluating the benefits and their realisation

By having a structured approach to benefits management, an organisation can ensure that the programme objectives contribute to the organisation's strategic objectives. In addition, it enables the capture of benefits not anticipated at the outset

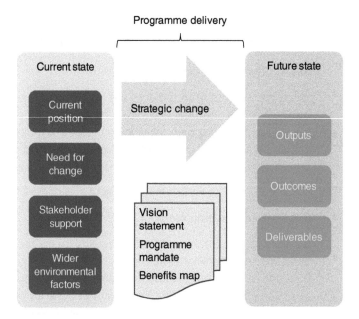

Figure 6.3 Managing benefits realisation.

of the programme and provides an ongoing focus, after the programme, on sustaining benefits.

The work to define the programme benefits, develop the programme delivery plan and define the project outputs needs to be integrated (see Figure 6.3). Work should start with gaining a better understanding of the benefits outlined in the programme brief and the business changes needed to deliver the benefits. This in turn provides the basis for the scope and the programme delivery plan and dictates what the projects need to produce as outputs. The process becomes iterative, as more details about benefits emerge in developing the programme definitions and the project outputs.

6.2.7 Benefits and disbenefits

Benefits

A benefit is (directly or indirectly) measurable improvement resulting from an outcome perceived as an advantage by one or more stakeholders that contributes towards one or more organisational strategic objective(s).

A benefit is a measurable improvement delivered by a programme that is seen by a stakeholder to be positive and worthwhile, for example, creating more green spaces in a particular borough or a perception via a residents' survey that the crime rate has gone down.

Types of benefits

Benefits can be classified in a number of ways. A distinction is made between financial benefits, which are measured in monetary terms, and non-financial benefits, which cannot be measured in monetary terms.

Financial benefits are further categorised as 'revenue', those benefits that give rise to immediate bankable returns, for example, capital receipts from the disposal of property, or 'non-revenue', for example, an efficiency gain leading to less time to complete a required task, which is a gain that cannot be converted into a reduction in staffing or cost.

Non-financial benefits include improvements across services and corporate functions (e.g. human resources, information and communication technologies and legal services) that can be measured using national and local non-financial performance indicators and the results of citizen and staff surveys.

Disbenefits

Disbenefits are the outcomes/outputs from a programme that are perceived by one or more stakeholders as negative, for example, new operational costs or loss of green space in an area due to the building of a new school. The same change can be seen by different stakeholders as both a benefit (net cost reduction through fewer staff) and a disbenefit (job losses). These disbenefits can be classified, managed and measured in the same way as benefits.

Disbenefits can be confused with risks, but whereas risks may be avoided, disbenefits will certainly be created by the programme and their impact must be managed. It is important to understand which stakeholder will lose out so that this can be managed. Furthermore, with proactive management, some disbenefits can be potentially turned into opportunities or, indeed, new benefits (for example, the creation of a new school may result in the loss of green space, but extended community use facilities may compensate for the loss).

6.2.8 Benefits identification and mapping

Often, a benefits map is created in a visual form to capture and communicate the benefits. The map can be used throughout the life of the programme to analyse any impacts on benefits caused by changes in programme direction or changes to the strategy as a whole.

There are many ways to map benefits. It is useful to build benefits maps in stages (e.g. start with a session to agree on the programme objectives) because the relationships between the various components can be complex, and it can be better to work on these outside sessions with stakeholders.

The steps outlined in the following sections provide an example of how this can be done – the example used in the illustrations considers a leisure facility transformation programme.

Step 1: Mapping programme objectives to strategic objectives

At the beginning, it will be necessary to agree on the programme's objectives if these are not already clear from the programme mandate (see Figure 6.4). The programme objectives are statements that explain what the programme sets out to deliver at the highest level in the context of strategic objectives.

It is important to ensure that the programme objectives are within the scope and power of the programme.

Step 2: Identifying and mapping benefits to programme objectives

In this step, the process involves identifying benefits and disbenefits. If a programme brief has been created, some benefits and disbenefits will have already been identified. In addition, more potential benefits may have come to light since the creation of the programme brief.

The identified benefits should then be mapped to the programme objectives (see Figure 6.5).

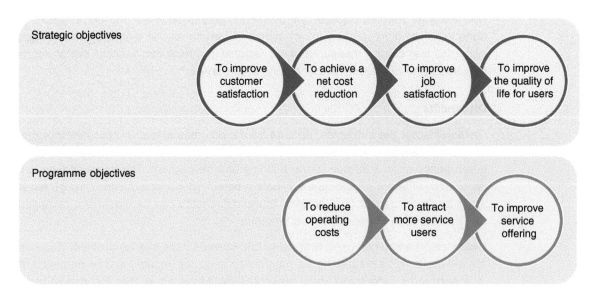

Figure 6.4 Benefits map Step 1: Mapping programme objectives to strategic objectives.

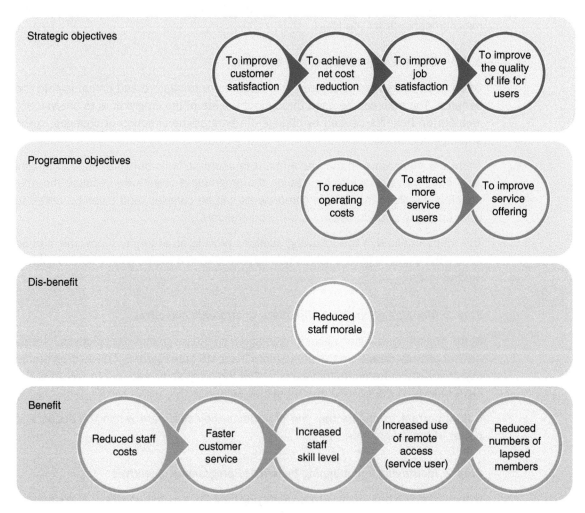

Figure 6.5 Benefits map Step 2: Identifying and mapping benefits to programme objectives.

6 Stage E: Benefits Realisation and Transition

Disbenefits are also shown to the right of programme objectives. It is also useful to indicate if there are any relationships between benefits and/or disbenefits by placing them adjacent to each other on the benefits map.

Benefits are often intertwined – if relationships between benefits start to become too complex, grouping can be used to identify the related benefits together, and the complexity can be captured in the benefits profiles. In scenarios where there are numerous links between the elements, links can be prioritised in order to maintain the visual clarity of the benefits map.

Step 3: Identifying business changes

The next step in the process is the identification of business changes that are needed in order to achieve the benefits (see Figure 6.6). Business changes are those made to the current ways of working that need to be implemented in the business areas

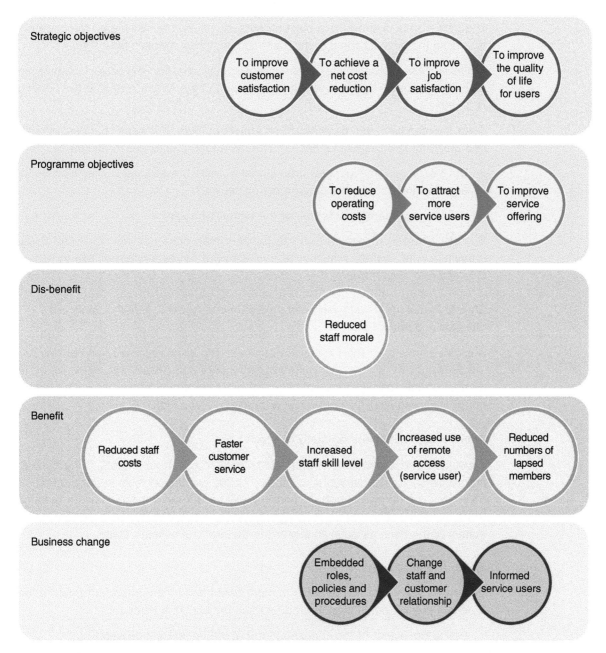

Figure 6.6 Benefits map Step 3: Identifying business changes.

affected by the programme. They can include process and behavioural changes and changes to operational procedures. For example, a new clear desk policy to support desk sharing will only deliver benefits if staff changes their habits and adequate storage is provided to staff.

Typically, the main business changes are noted on the map. A more detailed list of all business changes and how they will be implemented should be developed following the development of the benefits profiles. Business changes are numerous, and it would be difficult to try to fit them all on the map. It is, however, very valuable to use the map to consider all the major business changes required to deliver the benefits.

The activities needed to deliver the business changes should be included in the programme delivery plan (or separate benefits realisation plan) or where appropriate within the individual project plans.

Step 4: Mapping project outputs to benefits

The next step is to show the project outputs (enablers) that will create the capability to realise the benefits through the identified business changes (see Figure 6.7).

In the process of confirming, amending or adding benefits, the project outputs are reviewed to ensure that everything that needs to be created to enable the benefits is listed.

Once the map has been populated with project outputs, business changes, benefits and programme objectives, it can be used as a tool to consider the following:

* Do the programme benefits strongly align with the strategic objectives of the organisation through the programme objectives?

* Will the proposed projects deliver the benefits sought?

It is important to recognise that the benefits map does not indicate when things happen in time, that is, in sequence. It is simply a representation of how things are connected to each other.

Step 5: Mapping the links between programme objectives, benefits, business changes and project outputs

Once all elements of the map have been captured, links are identified between the enablers, business changes, benefits/disbenefits and programme objectives. The number of links can give an indication of the relevant importance of each element on the map and assist with benefits prioritisation (see Figure 6.8).

Prioritising benefits and business changes

It would not be appropriate or practical to track/measure all benefits. The benefits should be prioritised according to those that are critical to realising the programme objectives, those with important financial values and also those that are practical to measure (i.e. there is an existing baseline). In practice, this means that only a subset of the benefits is taken forward to the profiling benefits stage.

6.2.9 Profiling benefits

Benefits profiles describe benefits in more detail and record important information. A benefit profile should be prepared for each benefit. This helps to:

* Define the extent of the improvement that the benefit will deliver

* Ensure an appropriate person is accountable for the delivery of the benefits

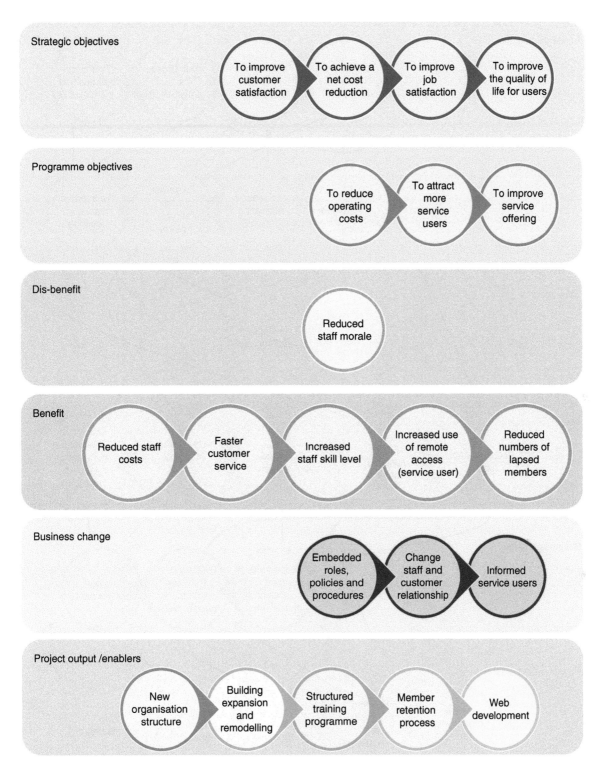

Figure 6.7 Benefits map Step 4: Mapping project outputs to benefits.

- Prioritise benefits
- Clarify the project outputs that are needed to enable the benefit to be realised

The details of benefits profiles will be refined and become clearer as the programme progresses.

See Appendix C for a benefits profile template.

6 Stage E: Benefits Realisation and Transition

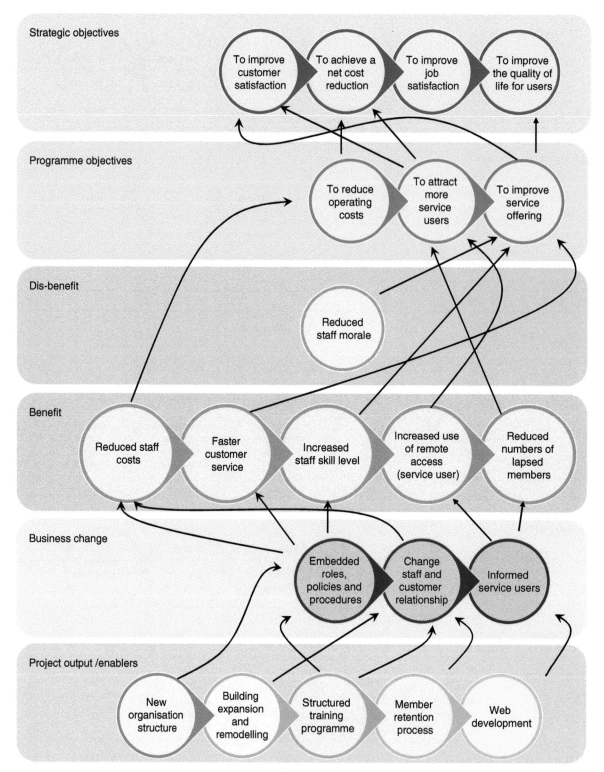

Figure 6.8 Benefits map Step 5: mapping the links between programme, objectives, benefits, business changes and project outputs.

6.2.10 Realising benefits

Many programmes fail to deliver their full potential because the responsibility and support are not put in place to deliver all the benefits. The biggest challenge is often to change the way people work.

There are two critical aspects to successful benefits realisation:

- establishing operational benefit owners, that is, people in the service/business who will be accountable for the achievement of the benefits

- implementing the business changes required to embed the new ways of working to deliver the benefits

The programme business change manager role has been developed in programme management to own and lead the realisation of benefits on a day-to-day basis. It is advisable to have an overall programme business change manager who is supported by the benefits owners in the service or business areas involved.

The programme business change manager should plan benefits realisation activities in conjunction with the benefits owners and service areas affected. The following activities should be considered:

- Business change activities, including decommissioning of old ways of doing things, for example, processes, systems, buildings, and posts, and implementing new ways of doing things, for example, new processes and training for the same

- Checkpoints for reviewing the changes and resultant benefits

- Responsibilities within the business change area

- Interdependencies and resources required

- Achievement of benefits that are 'early wins' and ways of maintaining focus on benefits that will take longer to achieve

The activities identified should be included in the PDP, included in a separate benefits realisation plan or, where appropriate, they may be included in individual project plans.

6.2.11 Reviewing and embedding

The programme business change manager is responsible for collecting information on the achievements of benefits, and where benefits are not on track, taking the necessary action to ensure they are realised.

There are many ways to do this: see Appendix C for an example of a template that can be used to record and track information on all the programme benefits. If required, the financial benefits information can be used for calculating overall financial benefits to date.

The information in a template such as this can also be aggregated with similar templates for other programmes to give a portfolio or strategic view of benefits realisation.

The purpose of tracking is to focus operational business units on achieving, sustaining and improving the benefits. This may mean using existing measures and reports (e.g. from the finance section). It may mean implementing new key performance indicators (e.g. number of lost calls in a call centre operation). In both cases, the measures need to be integrated into the operational management processes and systems, for example, reporting systems, management meetings, personal performance targets, etc.

By integrating benefits tracking into the existing reporting and management processes, benefits owners have the basis for refining and optimising benefits.

This involves looking at the trade-offs between benefits and the impact the operational environment is having on the level of achievement. For example, increasing the number of people working from home by introducing smart-working facilities may reduce office accommodation costs but also reduce effective communication and collaboration between team members.

It is also important to embed and reinforce the new ways of working and discourage the old.

During the life of the programme, benefits should be reviewed by the programme sponsor's board. Structures and mechanisms should also be established, which will continue the process of monitoring and reviewing achievements of benefits beyond the programme closure. The approach primarily relates to programming in a public sector environment (see Public Sector Programme Management Approach – detailed in Bibliography), but can be tailored to suit the specific context of a private sector programme as well.

6.2.12 Communications

The communication of the achievement of benefits is an obvious but often overlooked activity. It is important to communicate to stakeholders the success of the programme as it starts to achieve key benefits. Positive feedback and reinforcement are powerful ways to help people transition and adopt new ways of working.

A way of ensuring that the communication of success is not overlooked is to include in the communications and stakeholder plan actions around publicising benefits. Target dates for the achievement of benefits can be taken from the benefits profiles and used to identify when the communications activities should occur.

6.2.13 Sign-off

The key outputs from the benefits management and realisation activities are the benefits map and profiles and the benefits realisation plan (either a separate plan or included as part of the PDP).

The benefits map, profiles and programme plan are signed off by the programme sponsor during the definition phase.

The benefits profiles and the benefits realisation activities should first be agreed upon by the programme business change manager (and benefits owners if possible) before they are submitted for sign-off.

The benefits profiles and benefits realisation activities (and plan) need to be kept in alignment with the programme delivery plan, project outputs and changes to the objectives of the programme. These documents should be reviewed and changes signed off:

- Before starting a new part of the programme

- When there are major changes to the scope

- When there are developments that impact the attainment of the planned benefits

- Before transitioning to the new state (measure the performance of the current state)

- After transitioning to the new state (measure the performance of the altered state)

- At programme closure ensure that benefits ownership and realisation continue in business as usual to achieve the benefit targets set out by the programme

6.2.14 Transition strategy and management

Managing the transition strategy should be addressed from (i) an operational, (ii) programme organisation and (iii) business as usual point of view, as described below:

1. Operations: Programme close-out/operational readiness

   ▪ Assets and facilities ready for operations on time

   ▪ All facilities are fit for purpose

   ▪ Facilities operated for the first time with no critical failures

   ▪ Recognised legal, contractual and critical schedule requirements

   ▪ Comprehensive operations philosophy and plan in place

   ▪ Clear roles and responsibilities pre- and post-transfer

   ▪ Facilities easy to operate and maintain

   ▪ Staff to receive timely training to operate and maintain facilities

   ▪ Capture of lessons learned for knowledge dissemination

2. Organisation: Programme exit strategy

A key challenge is the management of resources to ensure that key programme staff remain for the life of the wind-up in critical functional areas to ensure that corporate knowledge is retained and there is clear accountability for any associated exit tasks. An interim resource plan using third-party resources could help to shore up unexpected staff losses and provide continuity.

A programme needs to consider how the transition to operations will impact the organisation and the implementation of an exit strategy as it reaches the end of its life (see Figure 6.9).

Key areas to consider and address:

- Governance

   ▪ Opportunity to formalise an exit strategy as a formal project/programme

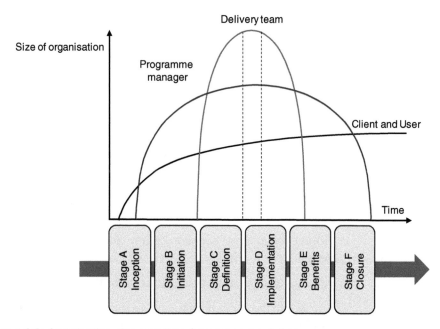

Figure 6.9 Organisation size over time for programme delivery.

- Resource management

  - Key programme staff leaves unexpectedly creating gaps in expertise and loss of corporate knowledge

- Commercial management

  - Clear visibility of potential contractual issues and tracking of any resulting claims/disputes using a formal system will help to make the exit as efficient as possible

- Legal

  - Developing a legal risk map will help highlight business requirements for organisational exit

- Stakeholder management and communications

  - Exit strategy would benefit from the development of a stakeholder map

- Risk management

  - Capturing and formalising will consider all exit strategy risks

- Information technology:

  - Information technology expertise may be required to help support legacy bodies

6.3 Key roles and responsibilities of this stage

6.3.1 Programme sponsor

- In conjunction with the programme manager, initiate the benefits realisation review

- Confirm with the programme business change manager and programme manager that the programme has achieved its objective and can be closed

- Determine if further work is required

- Determine the scope of any further works

- Determine how any additional work will be carried out

- Review with the programme business change manager the progress of transition arrangements

- Report to the programme sponsor's board on completeness of programme

6.3.2 Programme business change manager

- Together with the programme business realisation manager, carry out the final benefits realisation review

- Advise the programme sponsor that the programme outcomes that have been delivered will achieve the required benefits

- Advise programme sponsor of any additional work required to create further benefits

- Manage the transition arrangements for the new capability/enterprise

- Liaise with the client body to ensure their readiness to receive the new capability/enterprise

6.3.3 Programme business realisation manager

- Together with the programme business change manager, carry out the final benefits realisation review

- Determine if there is any benefits shortfall in the programme outcomes delivered

6.3.4 Programme manager

- Assist the programme sponsor/programme business change manager in the final benefits review

- Assist the programme sponsor in scoping any additional work required

6.3.5 Programme financial manager

- Advise the programme sponsor and programme manager on the financial implications of carrying out any additional work

6.3.6 Programme management office

- Advise the programme sponsor and programme manager of the schedule and cost implications of carrying out any additional work

6.3.7 Programme sponsor's board

- Review with the programme sponsor the conclusion of the final benefits realisation review

- Formally confirm the programme has achieved its objectives and can be closed

- Verify any additional works that need to be carried out

7 Stage F: Closure

7.1 Purpose of this stage

At the point at which the programme sponsor has agreed that the required outputs and the outcome defined in the programme delivery plan have been achieved and no further works are envisaged, the programme is considered complete.

Please refer to the Programme Delivery Matrix (PDM) in Appendix A to understand the key roles and artefacts that comprise the CIOB programme life cycle.

7.2 Key activities of this stage

The programme closure report is a key document for this stage (see Figure 7.1).

Key artefacts	CIOB code of practice programme management stages
Vision statement	Stage A - Inception
Programme mandate	Stage A - Inception
Programme brief	Stage B - Initiation
Programme business case	Stage B - Initiation
Programme delivery plan	Stage C - Definition
Programme progress report	Stage D - Implementation
Programme benefits report	Stage E - Benefits realisation and transition
Programme closure report	Stage F - Closure

Figure 7.1 Key artefacts: Stage F (Closure).

7.2.1 Shutting down the programme

Prior to disbanding the programme team, a number of final tasks need to be undertaken (see Figure 7.2):

- Lessons-learned review
- Archiving of programme/project information
- Transfer of building information modelling system to management of the new undertaking

Code of Practice for Programme Management in the Built Environment, Second Edition. The Chartered Institute of Building.
© 2024 John Wiley & Sons Ltd. Published 2024 by John Wiley & Sons Ltd.

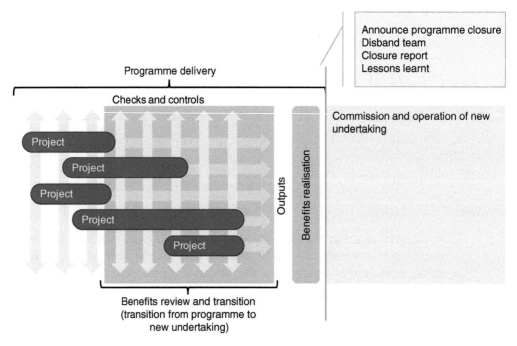

Figure 7.2 Stage F: Closure.

- Resolution of any contractual disputes and claims and settlement of final accounts with external parties

- Final programme communication advising stakeholders of closure

- Preparation of programme closure report and submission to programme sponsor's board

- Highlighting of any outstanding works or issues to programme sponsor's board

7.2.2 Disbanding the programme team

The manner of disbanding the programme team will vary depending on whether the staff are permanent employees of the sponsoring organisation or are externally contracted. For those who are permanent employees, a process of coordinating the timing of their release and locating other roles and opportunities needs to be implemented. Contracted staff leave the programme when the programme manager agrees that their role in the programme has been completed. It is likely the programme manager will stay for a period to liaise with and advise the team taking over the undertaking.

7.2.3 Programme closure

The programme closure stage is reached when the programme sponsor agrees that the required outcomes and outputs as defined in the programme delivery plan have been achieved and no further works are envisaged or when there is a significant change on one of the key parameters – for example, if there is a strategic change within the organisation that makes the business case no longer viable or the programme funding becomes no longer available.

It is necessary to ensure that a formal programme closure process is undertaken for all programmes, regardless of the reason for closure, so that there is a formal recognition that the programme has been completed and that delivery of capability and benefits has been assessed. This is particularly important for complex programmes or programmes that have a prolonged delivery period as the organisational 'drift' can otherwise set in, which may allow viewing the programme as part of normal business.

7.2.4 Preparing for closure

Typically, programme closure should happen after the completion of the last project or initiative. In deciding whether to close the programme, consideration should be given to whether the business case has been satisfied or is well on its way to being satisfied.

It is often the case, for large and complex programmes, to view programme closure as a separate project where a timeline for closure activities with specific tasks and ownerships is allocated. All relevant parties, including the key stakeholders, should be notified that the programme is being prepared for closure.

A closure activity plan should be prepared to include a programme review, formal sign-off and closure. The following questions should be considered as part of the plan:

* Are there any residual risks or issues that could affect operations?

* Are all the projects complete, including those that are designed to ensure embedding of the new operational environment?

* Has the programme delivery plan been achieved?

* Are all other programme activities complete?

* Has the business case been satisfied?

For programmes where closure is necessary due to a fundamental parameter change, confirmation should be sought that all projects have either been closed or, if their outcomes are still required, that alternative and appropriate governing arrangements have been put in place.

7.2.5 Programme review

Before closing the programme, a formal review should take place to assess whether it has delivered on what it set out to do. The programme sponsor should complete the review with input from the programme manager and programme business change manager. The following may be used to test if the programme has delivered against its objectives and outcomes:

* Vision

* Programme brief

* Benefits map

* Business case

* Programme delivery plan

Some programmes may not have been defined using these specific components. It is most important to review the capabilities that have been delivered and the outcomes and benefits so far.

The following should also be considered:

* Residual risks and issues – have these been assigned?

* Remaining transition activities – have these been allocated?

- Any support functions that require continuation, particularly if relevant to benefits realisation – have these been assigned?

- Any benefits not fully realised – have these been handed over to the relevant business area to be monitored?

- Lessons learned – are processes in place to capture and assimilate?

The programme review must assess the programme's performance and processes to identify and capture lessons learned, as these may benefit other programmes.

There may also be a need for an independent or external review of the overall programme delivery and performance prior to the formal closure.

7.2.6 Programme closure

Before recommending programme closure to the programme sponsor's board, the programme sponsor should be satisfied that the business case is delivered with all projects and programme activities complete. In addition, the programme sponsor should ensure that any required handover or transition activities have been defined and assigned to relevant business functions.

Programme closure should be authorised by the programme sponsor's board on the recommendation of the programme sponsor. Once the programme's closure activities are completed, the programme sponsor should then confirm programme closure to all relevant parties. To facilitate the governance requirements, often a closure report (see Appendix C for a template) will need to be prepared.

7.2.7 Communications

Communications at programme closure include ensuring that all achievements, that is, delivery of the PDP, outputs, outcomes and benefits, have been advised in an appropriate manner to stakeholders. Where possible, references should be made to the programme initiation documents, including the business case, so that, in simple terms, it can be clearly stated why the programme was initiated and what have been the final results and benefits.

7.2.8 Disband programme organisation and supporting functions

The programme organisation should be disbanded. This will include releasing all individuals and resources from the programme. Individuals may need to be redeployed back into the organisation, and this should be planned in advance. It is advisable to consider the new skills imparted on the individuals when reassigning them back into the organisation.

7.2.9 Sign-off

The programme sponsor's board authorises programme closure on the recommendation of the programme sponsor. If the programme sponsor's board is not satisfied with the recommendation, they should give clear direction as to further work required and ensure that resources are available to undertake the work as deemed necessary.

7.3 Key roles and responsibilities of this stage

7.3.1 Programme sponsor's board

- Instruct the programme sponsor to close down the programme

- Review closedown reports

- Identify any residual issues that need to be reported to the client

7.3.2 Programme sponsor

- Instruct programme manager to close down the programme

- Ensure all financial matters have been finalised

- Ensure any outstanding claims/disputes have been resolved

- Oversee preparation of a post-implementation review

- Oversee preparation of a lessons-learned log

- Ensure programme business change manager is providing appropriate transition support

- Receive confirmation from programme manager that programme has been shut down and the programme management team has been disbanded

- Ensure all stakeholders have been advised of programme closedown

- Ensure all relevant documentation has been transferred to client/new enterprise

- Confirm with programme business change manager any requirements for future benefits reviews

- Present closedown reports to programme sponsor's board

- Disband the programme sponsor's board

- Ensure any residual issues have been reported to the client

- Ensure any requirements for future benefits reviews have been reported to the client

7.3.3 Programme business change manager

- Assist in the preparation of the post-implementation report

- Identify lessons learned

- Determine any requirements for future benefits reviews

- Continue to liaise with client regarding the operation of the new capability or enterprise

7.3.4 Programme manager

- Advise the programme team of the termination of the programme

- Ensure all programme and project documentation is complete and archived

- Manage process of termination of involvement/employment of the programme team members

- Confirm process for notifying stakeholders of the termination of programme

- Resolve any outstanding claims/disputes with consultant, contractors and suppliers

- Ensure with programme financial manager all financial accounts are settled and closed

- Together with programme financial manager, prepare final cost statement

- Co-ordinate the preparation of the lessons-learned log

- Prepare programme post-implementation report

- Identify any outstanding or unresolved issues

- Disband the programme management board

7.3.5 Programme financial manager

- Together with programme manager, resolve any outstanding claims/disputes with consultant, contractors and suppliers

- Settle and close all financial accounts

- Assist in preparation of final cost statement

- Identify any lessons learned

7.3.6 Stakeholder/communications manager

- In conjunction with programme manager, advise all stakeholders of termination of programme

- Identify any lessons learned

7.3.7 Programme management office

- Ensure the programme and projects information system is complete and archived

- Assist programme manager in preparation of the post-implementation report

- Identify any outstanding or unresolved issues

- Identify any lessons learned

7.3.8 Health and safety manager

- Prepare final report summarising progress of health and safety matters during the programme

- Identify any lessons learned

7.3.9 Sustainability manager

- Prepare final report summarising progress of sustainability and environmental matters during the programme

- Identify any lessons learned

Appendix A
Programme Delivery Matrix

A.1 Background

The following matrix has been specifically developed for the 2nd edition to demonstrate how Stages A to F, described in this code of practice, relate to the RIBA plan of work 2022.[1] It is not an exact mapping but sufficient for readers to understand the key stages.

The main tasks described in this guide are itemised along with who is responsible for performing that task. The programme artefacts are shown, as described in this guide, and how they are constantly reviewed during each stage. In addition to this, 'supporting artefacts' as detailed within the RIBA Plan of Work are shown, which can supplement each of the programme stages.

A.2 Use of the Programme Delivery Matrix

This matrix should be used as a starting point for the further development and refinement of programme processes and procedures. Depending on the type of programme being delivered, the programme team may have to include supplementary artefacts to help in monitoring and controlling the programme.

---

[1] https://www.architecture.com/knowledge-and-resources/resources-landing-page/riba-plan-of-work

---

Programme Delivery Matrix

RIBA Plan of Work 2020

RIBA Plan of Work 2020 stage
[0] Strategic Definition
[1] Preparation and Briefing
[2] Concept
[3] Spatial
[4] Technical
[5] Manufacturing and Construction
[6] Handover
[7] Use

Supporting Artefacts (RIBA Plan of Work) and **Programme Artefacts (CIOB CoP)**

Supporting Artefacts (RIBA Plan of Work)	Programme Artefacts (CIOB CoP)	Associated stage
• Project Risks • Project Budget	• Terms of Reference	Stage A – Inception
• Site Information • Site Surveys • Project Execution Plan • Procurement Strategy • Responsibility Matrix • Information Requirements	• Benefits Appraisal • Programme Schedule	Stage B – Initiation
• Architectural Concept • Strategic Engineering • Cost Plan • Project Strategies • Design Programme • Change Control	• Benefits Profiles • Programme Schedule • Programme Work Breakdown Structure • Stakeholder Analysis • Risks, Issues, Assumptions & Constraints • Programme Controls • Programme Financial Plan • Programme Transition Plan	Stage C – Definition
• Site Logistics • Construction Programme • Site Queries • Commissioning • Building Manual	• Programme Progress Reports	Stage D – Implementation
• Plan for Use • Project Performance	• Programme Delivery Plan • Programme Benefits Profiles	Stage E – Benefits Realisation & Transition
• Final Certificate • Post Occupancy Evaluation • Project Outcomes • Sustainability Outcomes • Building Manual • Health & Safety	• Programme Lessons Learnt Log • Programme Delivery Plan	Stage F – Closure

Programme Delivery Matrix — Tasks (Plan / Do / Check / Act)

Task	Performed By	Stage	P/D/C/A
Prepare initial **vision statement**	client organisation	Stage A – Inception	Plan
Appoint programme sponsor	client organisation	Stage A – Inception	Do
Review **vision statement**	programme sponsor	Stage A – Inception	Check
Approve **vision statement**	client organisation	Stage A – Inception	Act
Develop **programme mandate**	programme sponsor	Stage A – Inception	Do
Appoint programme manager	programme sponsor	Stage A – Inception	Do
Appoint programme sponsor's board	client organisation	Stage A – Inception	Do
Appoint external consultant(s)	programme sponsor	Stage A – Inception	Do
Approve **programme mandate**	programme sponsor's board	Stage A – Inception	Act
Conduct stage gate review	programme sponsor	Stage A – Inception	Act
Allow the team to proceed to the next stage	programme sponsor's board	Stage A – Inception	Act
Develop the **programme brief**	programme manager	Stage B – Initiation	Do
Approve the **programme brief**	programme sponsor's board	Stage B – Initiation	Check
Appoint programme business change manager	programme manager	Stage B – Initiation	Check
Develop **programme business case**	programme manager	Stage B – Initiation	Plan/Do
Review **programme business case**	programme manager	Stage B – Initiation	Check
Approve **programme business case**	programme sponsor's board	Stage B – Initiation	Act
Conduct stage gate review	programme sponsor	Stage B – Initiation	Act
Allow the team to proceed to the next stage	programme sponsor's board	Stage B – Initiation	Act
Develop the **programme delivery plan**	programme manager	Stage C – Definition	Plan
Review **programme delivery plan**	programme manager	Stage C – Definition	Check
Approve **programme delivery plan**	programme sponsor's board	Stage C – Definition	Act
Appoint programme management board	programme sponsor	Stage C – Definition	Do
Appoint programme team members	programme manager	Stage C – Definition	Do
Establish programme management office	programme manager	Stage C – Definition	Do
Implement supplementary procedures	programme manager	Stage C – Definition	Do
Conduct stage gate review	programme sponsor	Stage C – Definition	Act
Allow the team to proceed to the next stage	programme sponsor's board	Stage C – Definition	Act
Appoint project management teams	programme manager	Stage D – Implementation	Do
Measure performance	programme manager	Stage D – Implementation	Check
Prepare **programme progress report**	programme management office manager	Stage D – Implementation	Do
Approve **programme management report**	programme manager	Stage D – Implementation	Check
Approve **programme progress report**	programme sponsor	Stage D – Implementation	Act
Conduct stage gate review(s)	programme sponsor	Stage D – Implementation	Act
Allow the team to proceed to the next stage	programme sponsor's board	Stage D – Implementation	Act
Undertake Programme Benefits Review	programme business change manager	Stage E – Benefits Realisation & Transition	Do
Prepare **programme benefits report**	programme business change manager	Stage E – Benefits Realisation & Transition	Do
Approve **programme benefits report**	programme sponsor	Stage E – Benefits Realisation & Transition	Check
Conduct stage gate review	programme sponsor	Stage E – Benefits Realisation & Transition	Act
Allow the team to proceed to the next stage	programme sponsor's board	Stage E – Benefits Realisation & Transition	Act
Confirmation of programme completion	programme sponsor	Stage F – Closure	Do
Conduct lessons learnt review	programme manager	Stage F – Closure	Do
Archive programme / project information	programme manager	Stage F – Closure	Do
Prepare **programme closure report**	programme manager	Stage F – Closure	Do
Approve **programme closure report**	programme sponsor	Stage F – Closure	Check
Conduct stage gate review	programme sponsor	Stage F – Closure	Act
Formally close the programme	programme sponsor's board	Stage F – Closure	Act

Appendix B
Tools and Techniques

B.1 Background

The following sections provide further information that programme teams could use to help them in effective delivery.

B.1.1 Stage review process

Each stage will be subjected to a stage review process. This will involve the programme sponsor and programme manager presenting to the programme sponsoring board the status of all the work done in the stage. It will highlight achievements, risks, issues, concerns and key decisions required (see Figure B.1).

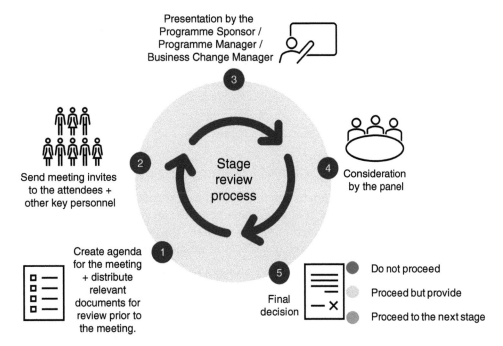

Figure B.1 Stage review process.

The program sponsoring board will review the documents presented to them and will provide authorisation to the programme sponsor to proceed to the next stage.

The UK Infrastructure and Projects Authority has developed an assurance review toolkit for public sector programmes. The documentation relating to gate reviews can be used as a basis for developing a framework for managing gate reviews (Infrastructure and Projects Authority, 2023).

B.1.2   Planning poker

In the earlier stages of the programme, the scope will be evolving, and the programme team may not be able to develop accurate forecasts for how long it will take to produce certain artefacts.

To ensure that the level of effort involved can be clearly communicated and agreed upon amongst team members, an agile technique used for IT projects can be utilised. This method of determining how much effort is likely to be required is called planning poker.

Planning Poker is based on the Fibonacci sequence.

The Fibonacci sequence starts at 0 and 1. The next number is the result of adding the previous 2 numbers, as shown in Figure B.2. The numbers in the Fibonacci sequence can be used as a rating scale to denote the difficulty of effort with 0 being the least and 21 being the most.

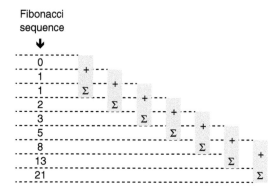

Figure B.2 Fibonacci sequence.

To utilise this method for construction programmes, consider the following scenario.

The programme team is tasked with producing the first draft of the programme business case. The programme sponsor can give a directive to the team that this needs to be produced in say two weeks. However, this is based on what the programme sponsor believes it will take to produce this document.

A more collaborative way would be to discuss this with the team that will be responsible for assembling the information for the document.

A facilitator is chosen from the team, and they will read out the details of the deliverable that needs to be completed. It could be something like:

> We need to produce the programme business case. This needs to be a good first draft and must include the initial programme costs as we understand them, the procurement method we are likely to use, the benefits, risks, assumptions, exclusions and any critical issues.

The team members are handed out cards with the following numbers: 1,2,3,5,8,13 and 21. The rating on the card relates to the story points. Story points are a measure of how much effort is required.

The team then thinks about what has been said by the facilitator and, when requested, they place a card face down on the table.

The number that has been placed face down represents the effort required to produce the document.

The facilitator then asks each person to reveal the number on their card. The facilitator will then ask the person with the highest score as to why they chose that number. That person would explain why they chose that number and their reasons for it.

A discussion around this will take place and the facilitator will then ask the players to re-submit their scores in round 2 (see Figure B.3).

After the second round, the scores are reviewed. From Figure B.3, it can be seen that there is a consensus that the team has a majority rating of 13 story points. If this is acceptable to the team, then this is the estimate for the number of days it will take to produce the programme business case, which can be agreed upon mutually.

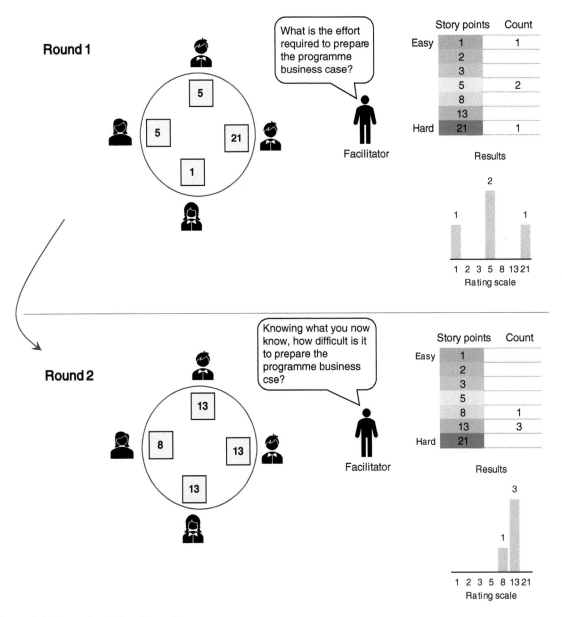

Figure B.3 Example of planning poker.

If no agreement was reached after round 2, the facilitator can ask for players to vote in another round until mutual agreement is reached.

This method does not replace the need to produce accurate estimates but merely is a tool for allowing the programme team to understand how much effort is required by them to produce artefacts.

It stops management from imposing time deadlines for producing artefacts/ deliverables and allows the team to assess how much time they need to do a good job.

B.1.3 Managing work products using agile methods

Agile development programmes in the IT industry tend to use time-boxing or sprints. These are intense periods of work followed by a review or retrospective at the end of the cycle. Daily stand-up meetings are held during the sprint to ensure that the team has all the help they need to finish the work.

The daily stand-up meetings are not technical meetings. They deliberately last between 10 and 15 minutes and focus on the following:

● What work are you doing?

● Do you need help in completing your task?

● Are there any issues?

This method relies on the energy and inertia created by experienced team members to ensure that the final product is the most optimal solution (see Figure B.4).

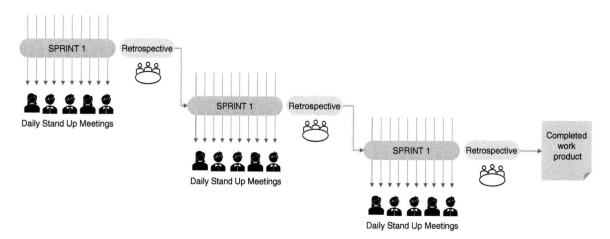

Figure B.4 Manging using the concept of SPRINTS.

The nature of construction programmes is different from those in the IT industry; however, use can be made of these techniques in the early stages of programmes to ensure that the right approach is chosen.

B.1.4 Managing the time schedule

Each stage will involve key decisions that need to be made and important tasks that need to be completed. A sensible approach to recording, tracking and managing the time schedule needs to be ascertained.

Having a clear time schedule can help ensure that important tasks are not missed and also form the basis of a communication tool for the programme sponsor and the key stakeholders.

At the earlier stages of a programme, the detail may be an evolving process, and therefore a light-tough approach could be adopted until the detail is worked out in the subsequent stages.

A Kanban board could be used for communicating the key tasks to the team, and this can be supplemented by a Gantt chart-style time schedule later on (see Figure B.5).

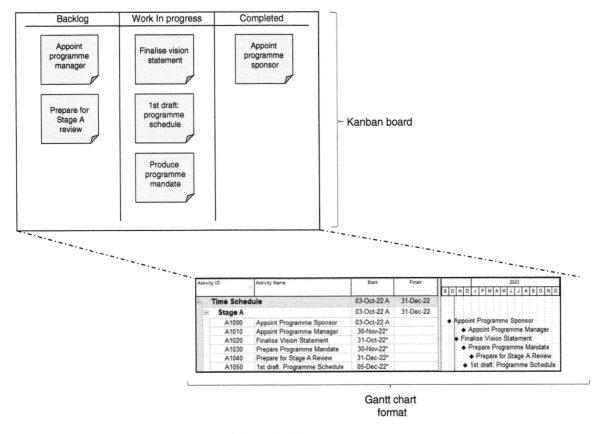

Figure B.5 Example of Kanban boards and time schedules.

The programme time schedule will develop over the course of the programme and will capture the work in the form of activities, as and when they are determined (see Figure B.6).

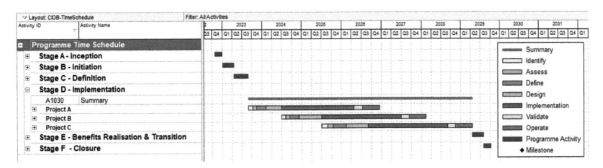

Figure B.6 Programme time schedule.

Is it important that all work activities are kept on the same schedule and that different views are adopted by the programme team for communication purposes (see Figure B.9).

B.1.5 Burndown charts

Another agile reporting tool that can be used in the early stages of the programme is the burn-down chart (see Figure B.7).

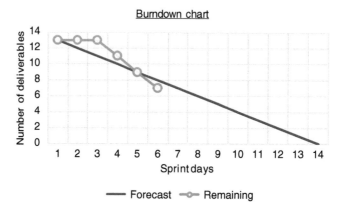

Work Products/Items	Estimate	1 2 3 4 5 6 7 8 9 10 11 12 13 14	Forecast	Remaining
Vision statement	3	● ● ●	13	13
Programme Mandate	5	● ● ● ● ●	12	13
Risk Register	3	● ● ●	11	13
Assumptions Register	3	● ● ●	10	11
Change Control Process	5	● ● ● ● ●	9	9
Stage Review Process	5	● ● ● ● ●	8	7
Programme Time Schedule	5	● ● ● ● ●	7	
Stage A - Time Schedule	5	● ● ● ● ●	6	
Process & Procedures	4	● ● ● ●	5	
Terms of Reference	3	● ● ●	4	
Meeting Structure	3	● ● ●	3	
Communication Matrix	3	● ● ●	2	
Stakeholder Analysis	5	● ● ● ● ●	1	
13	**52**		0	

Burndown chart

Figure B.7 Example of burn-down charts.

This could be used to monitor the number of artefacts/work products that need to be completed.

One important point to note is that if a task is given a three-day duration, it does not mean that it has to be completed in that time. This is an estimate, and it could be completed sooner.

B.1.6 Media wall

To ensure that communication is clear, a media wall should be decided upon for the programme team. This will provide high-level information for all the relevant team members and stakeholders.

It is quick and easy to use and eliminates the need to write up reports. It is designed in such a way that it can be updated manually as and when progress is made (see Figure B.8).

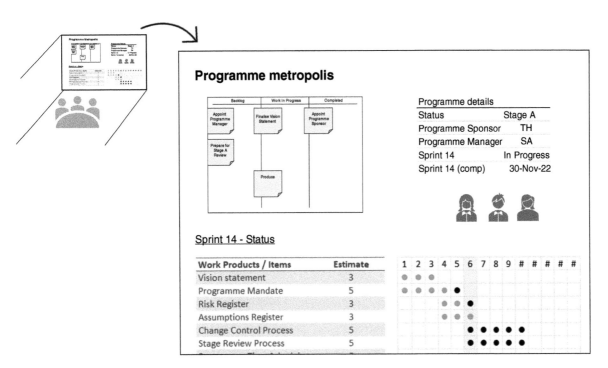

Figure B.8 Media wall example.

B.1.7 Managing the programme scope

The programme team must monitor and track the programme scope as and when further clarity is developed.

The scope of the programme will provide the basis for work/deliverables, which can then be scheduled accordingly. Once the time schedule is developed, other supporting processes can be blended into programme delivery as deemed necessary (see Figure B.9).

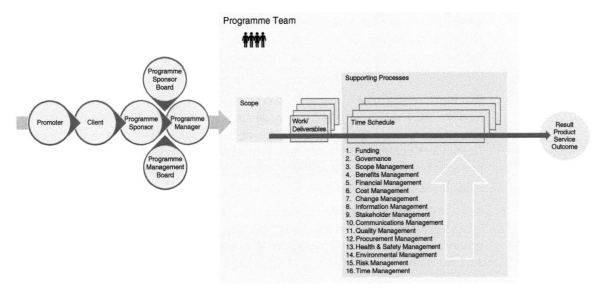

Figure B.9 Managing the programme scope.

B.1.8 Three-point estimating

The following areas will need to be considered during this process:

- Accuracy of data and assumptions

- Inconsistencies in data set

- Data errors

- Need for a wide range of perspectives

- Consideration of a risk mitigation plan that can limit risk impact

Given the number of estimates and activities on a programme, the three-point estimate (see Figure B.6) can sometimes be misleading. Another estimating technique is to enter variances of the probability distribution around the most likely estimate. This technique is based on an integrated holistic understanding and current knowledge of the programme's inherent risks. The estimation of uncertainty is illustrated in Figure B.10.

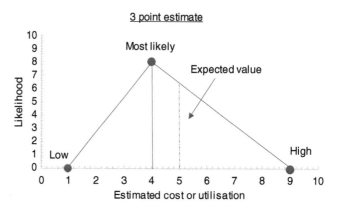

Figure B.10 Three-point estimate triangle.

A QRA undertaken on the programme will confirm the appropriate level of contingency (also known as 'overall risk pot') required to deliver the programme and be shared between the funding organisation, programme and projects. This includes the assessment of risks within individual projects in addition to cost and schedule risks across the programme.

The basis of the risk model should be an agreement on risk allocation between funders. The level of contingency proposed in the budget should represent those risks agreed to be under the influence and control of funders, programme and projects. The funder's contingency relates to the risk that does not sit with the programme and projects. The programme contingency relates to risks that do not sit within individual projects and are not covered by project contingencies.

In summary, an s-curve will represent the likely exposure of the total risk across the programme. Based on the agreed allocation of risk (at the 80th percentile), the contingency requirement, inclusive of VAT, will be estimated – that is, there is an 80% likelihood of not exceeding this contingency. The curve will also indicate that, from the analysis, an additional £xxxm would be required to secure confidence at the 95% level.

The s-curve will indicate an upper limit that is significantly greater than the 80th percentile; to some extent, this is influenced by the probabilistic model and the potential for uncontrollable acceleration costs or design risks. This suggests an increased maximum out-turn cost, albeit at a low level of probability. It is important that funders recognise increased exposure throughout the programme as it represents a substantially higher potential cost above the level at which the programme contingency is calculated (see Figure B.11 and Figure B.12).

Ref	Classification	Uncertainty	Overrun
A	Routine, been done before	Low	0% to 2%
B	Routine, but possible difficulties	Medium to low	2% to 5%
C	Development, with little technical difficulty	Medium	5% to 10%
D	Development, but some technical difficulty	Medium to high	10% to 15%
E	Significant effort, technical challenge	High	15% to 25%
F	No experience in this area	Very high	25% to 50%

Note: *The % range for uncertainty will depend on the appetite of the organisation

Figure B.11 Estimation of uncertainty: illustrative example.

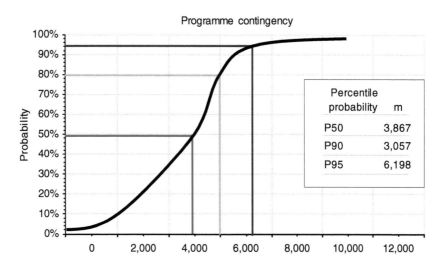

Figure B.12 S-curve detailing the cumulative contingency requirement.

There are other ways to represent risk and probability graphically as tornado diagrams or bar charts using Monte Carlo analysis and proprietary software for cost and time probability assessment.

It is apparent from programme delivery history that the risk of strategic misrepresentation remains high for complex programmes. Senior managers will need to consider the level of optimism bias (the tendency to overestimate the achievability of planned actions) for any programme. Benchmarking, due diligence and historical local performance analysis will allow the programme to avoid blind spots and prepare the organisation for black swan events.

The term 'black swan' – popularized by Nassim Nicholas Taleb – refers to unpredictable events with often have catastrophic consequences.

B.1.9 Reference class forecasting – example

Scenario

Your programme has information on 10 previously completed projects that all had similar scope. You would like to see what time allowance should be allowed for in the programme forecasts. The head of the programme office can provide information on these projects (see Figure B.13).

Dataset

Ref	duration (days)
Project A	300
Project B	350
Project C	375
Project D	350
Project E	365
Project F	380
Project G	350
Project H	320
Project I	390
Project J	400

Figure B.13 Dataset example.

Solving RCF using the manual method

Reorder and sort the durations in ascending order starting with the lowest value. Then divide the rows so that they all add up to 100%. Once this is done, work out the cumulative percentages (see Figure B.14).

Ref	duration (days)		cumulative
Project A	300	10%	10%
Project B	350	10%	20%
Project C	375	10%	30%
Project D	350	10%	40%
Project E	365	10%	50%
Project F	380	10%	60%
Project G	350	10%	70%
Project H	320	10%	80%
Project I	390	10%	90%
Project J	400	10%	100%
		100%	

Figure B.14 Reference Class Forecasting using the manual method.

Now we have enough information to manually plot the graph with cumulative percentages on the x-axis and durations on the y-axis (see Figure B.15).

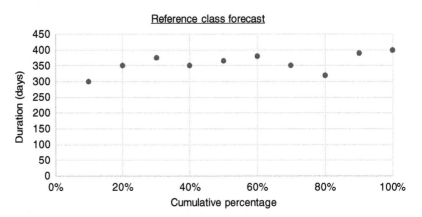

Figure B.15 Results of the manual method.

Solving RCF using excel

Looking at the dataset provided, ascertain the descriptive statistics as shown in Figure B.16.

Descriptive statistics	
Min	300
Mean	358
Max	400
Median	357.5
Mode	350

Figure B.16 Descriptive statistics (example).

We can now plot the frequency histogram using the bins sizes shown in Figure B.17.

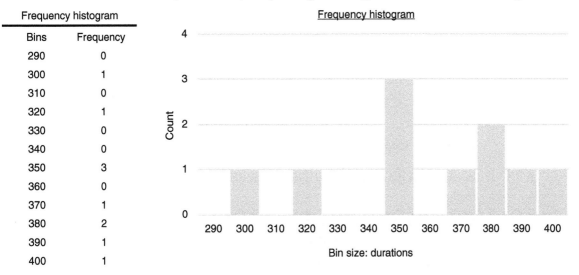

Frequency histogram

Bins	Frequency
290	0
300	1
310	0
320	1
330	0
340	0
350	3
360	0
370	1
380	2
390	1
400	1

Figure B.17 Frequency histogram (example).

We can see that the highest occurring is that of 350 days from the data set.

Now, using the PERCENTILE.INC function in excel we can produce the data for the reference class forecast (see Figure B.18).

P Value	Percentile	
P 95	5%	309
P 90	10%	318
P 85	15%	331
P 80	20%	344
P 75	25%	350
P 70	30%	350
P 65	35%	350
P 60	40%	350
P 55	45%	351
P 50	50%	358
P 45	55%	364
P 40	60%	369
P 35	65%	374
P 30	70%	377
P 25	75%	379
P 20	80%	382
P 15	85%	387
P 10	90%	391
P 5	95%	396

Figure B.18 Percentiles calculation using excel.

From this data, we can also produce the reference class forecast as shown in Figure B.19.

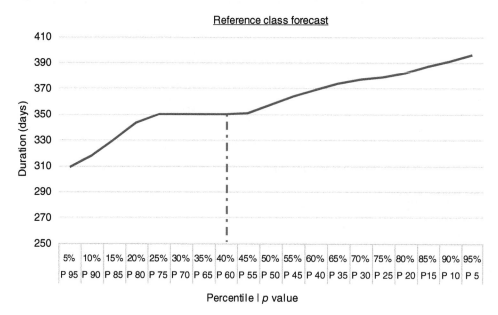

Figure B.19 Reference class forecast (example).

We can see that 40% of the data indicates that the project duration of 350 days is likely.

In other words, we can say that there is a P60 value of 60% change and that the project will take 350 days to complete.

This forecast could be used as a comparator to other forecasts produced on the programme.

Appendix C
Templates

Templates are based on the CIOB Code of Practice for Project Management, 5th edition and the public sector programme management approach. The following templates have been included in this document:

1. Vision statement

2. Programme mandate

3. Programme brief

4. Programme business case

5. Programme highlight report

6. Programme benefits profile

7. Programme closure report

C.1   Vision statement

Name of the Programme:
Vision Statement:
Name of the Programme Sponsor:

Version	Date	Status	Author	Change Description

Approved for submission to programme sponsor's board		Date	
Programme sponsor's board sign off to proceed		Date	

Distribution List		
Name	**Organisation**	**Role/Function**

Code of Practice for Programme Management in the Built Environment, Second Edition. The Chartered Institute of Building.
© 2024 John Wiley & Sons Ltd. Published 2024 by John Wiley & Sons Ltd.

C.1.1 Overview

Outline the corporate aspirations, setting out the business intent and benefits being sought.

C.1.2 Business context

Present overview of the state as it is now.

C.1.3 Strategic need

What are the business needs that the organisation is aiming to achieve? The needs should be clear and specific, not generic aspirations.

C.1.4 Programme vision

How would the future look when the programme is delivered and the benefits are achieved? This can be represented quantitatively and/or qualitatively.

C.1.5 Constraints and limitations

Include known and foreseeable constraints or exclusions that may apply and any other outcome that will be necessary but not within the programme.

C.2 High-level programme scope

Provide an overview of the scope of the programme, outputs and deliverables and list of projects, if possible.

C.3 Programme mandate

Name of the Programme:
Programme Mandate:
Name of the Originator:

Version	Date	Status	Author	Change Description

Approved for submission to programme sponsor's board		Date	
Programme sponsor's board sign off to proceed		Date	

Distribution List

Name	Organisation	Role/Function

C.3.1 Business need/vision statement

Use this section to set out the drivers that have created the need for this programme. This will include how the programme contributes to the organisation's strategic objectives and fits with other initiatives.

C.3.2 Outcomes

Briefly articulate the outcomes that the programme is expected to achieve. Specify if there are any constraints (e.g. must be achieved by x).

C.3.3 Next steps

List the activities, time and resources required to complete the programme brief and programme delivery plan.

Activity	Time	Resources	Costs

C.3.4 Sign-off

The programme mandate needs to be signed off by the sponsoring board, which will commit the resources to develop the programme brief and programme delivery plan.

C.4 Programme brief

Name of the Programme:
Programme Brief:
Name of the Originator:

Version	Date	Status	Author	Change Description

Approved for submission to programme sponsor's board		Date	
Programme sponsor's board sign off to proceed		Date	

Distribution List

Name	Organisation	Role/Function

C Templates

C.4.1 Programme vision

Describe a compelling picture of the future that this programme will enable. This should include the new and/or improved services and how they will look, feel and be experienced in the future.

C.4.2 Financial benefits

Describe the measurable improvements that the programme will achieve.

Benefit Description	Current Value	Target Value	Timing	Cashable Value	Non-Cashable Value	Benefit Owner

C.4.3 Non-financial benefits

Describe other benefits that will arise from this programme that are not easily measured.

C.4.4 Disbenefits

Describe the negative results of undertaking this programme.

C.4.5 Programme activity and projects

Describe the project and programme activities identified so far that will be required to deliver the programme benefits, with estimates of what they will cost and how long it will take to complete the work.

Programme Activities

Programme Activity	Description/Output	Duration	Costs	Lead Person
Benefits mapping exercise	Benefits maps Benefits profiles	6 weeks	£2500	A. Smith

Projects

Project	Description/Output	Duration	Contribution to Benefits	Costs	Lead Person

C.4.6 Quick wins

State what business activities should start, be done differently or cease in order to achieve quick wins.

C.4.7 Key risks and issues

List the potential threats (risks) and current issues to the benefits of the programme as they are currently understood. If there is one, use the corporate approach to risk and issue management. This section should be structured according to corporate guidance.

Risks – Anticipated threats to the benefits						

Description	Likelihood	Impact	Proximity (when it is likely to occur)	Risk Owner	Mitigating Action	Action Owner

Issues – Current threats to the benefits				

Description	Priority	Issue Owner	Action	Action Owner

C.4.8 Financial information

Set out the estimated financial costs and benefits.

List all currently identified or potential sources of funding.

Describe how these figures in the tables below have been arrived at, outlining all your assumptions.

C.4.9 Financial costs

Estimated Financial Costs – Capital (£000s)					
Description	Year 1	Year 2	Year X	Total	Ongoing
Totals					

Estimated Financial Costs – Revenue (£000s)					
Description	Year 1	Year 2	Year X	Total	Ongoing
Totals					
Total Costs – Capital and Revenue					

C.4.10 Constraints

Describe any known constraints that apply to the programme.

C.4.11 Assumptions

Describe any assumptions made that underpin the justification for the programme.

C.4.12 Programme capability

Describe how the organisation will provide the necessary programme management resources and capability required to carry out the proposed programme successfully.

C.4.13 Sign-off

This section should be signed by a representative of the sponsoring group to confirm acceptance of the brief. Use the version and authority sign-off on the front page.

C Templates

C.5   Business case

Name of the Programme:
Business Case:
Name of the Programme Manager:

Version	Date	Status	Author	Change Description

Approved for submission to programme sponsor's board		Date	
Programme sponsor's board sign off to proceed		Date	

Distribution List

Name	Organisation	Role/Function

This document provides a template for a business case in support of the investment decision.

In some cases, an outline business case may have been completed and agreed upon prior to the submission of this document for approval.

The main purpose of the business case is to provide evidence that the most economically advantageous offer is being pursued and that it is affordable. In addition, the business case explains the fundamentals of the programme and demonstrates that the required outputs and deliverables can be successfully achieved.

For small-scale programmes, a business justification document may be prepared to support the investment decision.[1]

C.5.1   Contents

- Executive summary
- Strategic case
- Economic case
- Commercial case
- Financial case
- Management case

C.5.2   Appendices

- Economic appraisals

---

[1]   See https://www.gov.uk/government/publications/the-green-book-appraisal-and-evaluation-in-central-government for further information and guidance.

- Financial appraisals

- Benefits register

- Risk register

- Stakeholder support assessment

- Strategic business plans

- Proposed delivery plan

- Change management plans

C.5.3 Executive summary

An executive summary indicates what decision is being sought, and what is the basis of the recommendation.

C.5.4 Strategic case

The strategic case summarises the vision and the strategic drivers for this investment, with particular reference to supporting strategies, programmes and plans.

The strategic context

Contents of this section may include the following:

- Organisational overview

- Current business strategies

- Other organisational strategies

The case for change

The case for change summarises the business needs for this investment, with particular reference to existing difficulties and the need for service improvement. This should clearly set out the investment objectives and the related benefits aspirations. Contents may include these elements:

- Existing arrangements

- Business needs

- Potential business scope and key operational requirements

- Investment objectives

- Main benefits criteria

- Main risks with the controls proposed

- Constraints and dependencies

C.5.5 Economic case

The economic case should include the options appraised and the outcomes, including the critical success factors. Contents may include the following:

- Critical success factors

- Long and short list of options

- Economic appraisal of options

- Estimated benefits

- Estimated costs

- Qualitative benefits appraisal

- Qualitative risk appraisal

- Sensitivity appraisal including scenario considerations

- Overall conclusion and recommendation

C.5.6 Commercial case

A commercial case may include the delivery model, risks and contingency options. Appropriate indices may be used to represent non-quantifiable risks and benefits. The contents may include the following:

- Agreed outputs and deliverables

- Risk management mechanism including risk transfer arrangements

- Contractual arrangements (including risks and personnel issues)

- Delivery timescales

- Accountancy arrangements

C.5.7 Financial case

The financial case should demonstrate the affordability model, where the scheme requires the support and approval of external parties, and indicate that this is committed and forthcoming. A letter of support should be attached as an appendix. The contents of this section may include the following:

- Impact on the organisation's income and expenditure profile

- Funding and expenditure profile

- Affordability and balance sheet treatment

C.5.8 Management case

The management case should include the management arrangements, including delivery, benefits realisation and risk management. Contents of this section may include these elements:

- Programme management arrangements

- Project management arrangements

- Use of specialist advisors and consultants

- Arrangements for change management

- Arrangements for benefits realisation

- Arrangements for transition management

- Arrangements for risk management

- Arrangements for delivery governance and review

- Contingency plans

C.6 Monthly programme report

Anticipated Final Cost (AFC) by WBS							
WBS	Cost Item (all costs in £m)	Baseline	Earned Value to Date	Earned Value for Period	Approved Changes	Anticipated Final Cost	Variance
1.1.1	A						
1.1.2	B						
1.1.3	C						
1.1.4	B						
1.1.5	A						
1.1.6	B						
1.1.7	C						
1.1.8	Indirect costs						
	Total						
1.1.9	Contingency						
	Total						

C.7 Programme highlight report

Name of the Programme:
Programme Highlight Report No. X:
Name of the Programme Manager:

Version	Date	Status	Author	Change Description

Approved for submission to programme sponsor's board		Date	
Programme sponsor's board sign off to proceed		Date	

Distribution List		
Name	Organisation	Role/Function

Programme Highlight Report		
Report Number	Date of Report	Period (From – To)

Programme Name:		Ref:	
Programme Vision:			
Programme Sponsor:		Telephone	E-mail
Programme Manager:		Telephone	E-mail

Overall Programme Progress and Status

	RAG Status		Comment on overall progress, status and any recommended actions
	This Period	Last Period	
Time			
Cost			
Delivery/ Outcome/ Output			
Benefits			

1. Overall Programme Financial Overview

Expenditure Type	Total Budget Amount	Total Forecast Spend	Variance Against Budget	%	Spend to Date
Capital					
Revenue					

Comment on financial position and any recommended actions

2. Progress this period

Projects	Detail

3. Milestones Overdue

Project/Sub-Programme	Milestone Description	Expected End Date	Revised End Date	Dependant Tasks/ Milestones? Y/N	Owner

4. Escalated Issues (including those from the last highlight report not yet resolved)

Item	Issue	Recommended Action(s)	Owner
4.1			
4.2			
4.3			

5. Escalated Risks

Item	Risk	H,M,L	Recommended Action(s)	Owner
5.1				
5.2				
5.3				

6. Milestones/Actions for Next Period (in addition to those overdue)				
Item	**Projects**	**Activity**	**Due Date**	**Owner**
6.1				
6.2				
6.3				
6.4				
6.5				
6.6				
6.7				

C.7.1 Programme Detail

Performance Against Plans								Financial Performance					Comment
		RAG Status											
		Time		Cost		Delivery							
Project/Sub-Programme	Specific ID	This Period	Last Period	This Period	Last Period	This Period	Last Period	Expenditure Type	Total Budget Amount	Total Forecast Spend	Variance Against Budget	%	
		G		G		G		Capital					
								Revenue					
		G		G		G		Capital					
								Revenue					
		G		G		G		Capital					
								Revenue					
		G		G		G		Capital					
								Revenue					
		G		G		G		Capital					
								Revenue					
		G		G		G		Capital					
								Revenue					
		G		G		G		Capital					
								Revenue					

Explanation of RAG Status:

Red – Overall slippage
Amber – Slippage in current period
Green – As planned
Grey – Slippage previously reported, recoverable within overall programme
Black – Project completed, on hold or cancelled

Note: The distinction between capital and revenue is only relevant to public sector programmes

C.8 Benefits profile

Prepare a benefits profile for each benefit. Benefits profiles describe benefits in more detail and record information to:

- Define the extent of the improvement that the benefit will deliver
- Ensure an appropriate person is accountable for the delivery of the benefit
- Prioritise benefits
- Clarify the project outputs that are needed to enable the benefit

Benefit profile:			
Benefit description	*Summary of benefit*		
Benefit type	*What are the financial (revenue or non-revenue), non-financial or disbenefit?*		
Programme business changes required	*What are the operational changes needed to achieve the benefits?*		
Outputs contributing to this benefit	*What are the activities or projects that, together with the business changes, will create the capability to realise the benefits?*		
Benefit owner	*Who will be responsible for making sure that this benefit is realised?*		
Stakeholder beneficiary	*Which stakeholders benefit from this improvement (or in the case of a disbenefit, which will be affected?)*		
Measurement and costs	*How will you know that the benefit has been achieved – what measures will you use? This could be existing performance indicators.*		
Dependencies on other programmes	*Are other programmes involved in helping to realise this benefit?*		
Assumptions	*Are there any assumptions that have been made that underpin the realisation of this benefit?*		
Constraints	*Are there any constraints that restrict the level of benefit that can be achieved?*		
Risks to benefit	*What are the risks that could prevent this benefit from being realised?*		
Measures	Baseline	Target (s)	Measurement method and responsibility
Description of measure	*Starting point from which you will measure this benefit*	*Target value and timescale (there could be several interim targets until you achieve the final target)*	*How the benefit information will be captured and who is responsible*

Benefit profile:			

C.8.1 Sign-off

The benefits profiles should be signed off by the programme sponsor (or programme business change manager on their behalf) to confirm acceptance of the benefits profiles.

C Templates

C.9   Tracking benefits: benefits-monitoring

	Benefits ID/name	Owner	Baseline	Target Value	Target Date	Last Review and Achievement to Date	Next Review	Action Points with Ownership
Financial revenue								
Financial Non-revenue								
Non-financial								

C.10   Programme closure report

Name of the programme:

Programme closure report:

Name of the programme manager:

Version	Date	Status	Author	Change Description

Approved for submission to programme sponsor's board		Date	
Programme sponsor's board sign off to proceed		Date	

Distribution List		

Name	Organisation	Role/Function

C.10.1   Context

Use this section to outline the history of the programme, including a few paragraphs that include why it was needed, when it started, and the reason for its closure.

C.10.2   Delivery

Outline how much of the programme has been implemented. If not all of its components have been created, outline the reasons for the shortfall; if the partially changed state of the organisation requires any further activities, where does the responsibility for those activities now lie.

C.10.3 Benefits

Using the table below, for each benefit list, the benefit measures that have been captured to show how well the business case has been achieved. The first three columns are copied from the benefits profile.

Benefit Description	Target Value	Achieved Value	Target Date	Achieved Date

C.10.4 Handover

For any benefit that has yet to be fully realised, list who is the new owner of the benefits realisation activities and whether this responsibility has been formally handed over and accepted.

C.10.5 Risks

List all outstanding risks and where the ownership now lies.

Risk	Previous Owner	New Owner

C.10.6 Issues

List all outstanding issues and where the ownership now lies.

Issue	Previous Owner	New Owner

C.10.7 Projects (optional section for premature closure only)

Normally, all projects will be closed at the end of the programme. In the case of premature closure, there may be some projects that will still be valuable. List the existing projects that will still be useful to the organisation and where the new ownership now lies.

Project name	Reason for Continuing	New Owner

C.10.8 Lessons learned

Highlight key lessons learned (positive and negative) that should be passed on to ongoing and future programmes. Consider the following:

- Governance organisation
- Stakeholder engagement and communications
- Vision and blueprint creation and delivery
- Benefits realisation

- Business case management

- Financial management

- Resource management

- Programme risk and issue management

- Programme planning, monitoring and control

- Quality management

- Change control

C.10.9 Sign-off

The closure review needs to be signed off by the programme sponsor, who will report to the programme sponsor's board in order to gain approval for the formal closure of the programme.

Use the version and authority sign-off on the front page.

C Templates

Appendix D
Key Roles: Skills and Competencies

The skills, competence and key criteria for the roles outlined are for guidance purposes only. The specifics will vary depending on the individual requirements and context of each role and programme.

D.1 Programme manager

D.1.1 Main duties

1. Lead and direct the programme management team, comprising:

   a. Planning and control
   b. Cost
   c. Finance
   d. Risk and opportunity

2. Other programme support as required from time to time such that the programme team successfully provides the required support, guidance, analysis and advice at the project and programme level as required

3. Building and maintaining strong relationships within the client organisation to ensure that the programme team integrates with the client organisation, addressing any cultural and procedural issues that may arise with integration in a positive and constructive manner

4. Implementation of a structured programme management methodology, including supporting processes, procedures and tools

5. Analysis of information from projects at programme level, with outcomes reflected in period reporting to the programme sponsor. Analysis to reflect the sensitivity of information and to include recommendations for actions, covering cost, funding, risk and opportunity, time, quality, interdependency, etc.

6. Continually seek to identify and fulfil client requirements and meet them in innovative and structured ways that add value and increase the probability of programme's success

7. Positive promotion of the programme to all key stakeholders

Code of Practice for Programme Management in the Built Environment, Second Edition. The Chartered Institute of Building. © 2024 John Wiley & Sons Ltd. Published 2024 by John Wiley & Sons Ltd.

D.1.2 Key competencies

Leading others

1. Works across boundaries, sharing information and matching resources to priorities

2. Is a visible leader who inspires trust, actively uses teamwork to deliver objectives and takes responsibility for overcoming setbacks

3. Communicates at all levels and with diverse groups and is able to present complex information clearly

4. Is honest and realistic, providing clear direction and focusing on strategic outcomes

Managing people and performance

1. Communicate and agree on measurable objectives with teams and staff

2. Manage change and continuous improvement dealing with resistance and conflict in a constructive way

Project and programme management

1. Make cross-cutting connections between issues and departments

2. Use communication strategies to present ideas in a clear and positive way

3. Be aware of the wider political environment

4. Analyse and use evidence

5. Use evidence to evaluate projects and programmes

6. Engage with relevant specialists to supply and evaluate all evidence

Financial management

1. Ensure that agreed objectives are delivered on time and within budget

2. Interpret trends and risks in financial management reports

3. Understand the wider expenditure and financial decision-making environment

4. Set targets to improve value achieved from resources

D.1.3 Key criteria

Key communicator recognises that the role of programme manager demands regular contact and negotiation with the team, clients, consultants, contractors and other stakeholders. Focus on customer satisfaction is paramount.

An essential skill is being able to see the big picture, recognise the details of projects and how they could contribute to the success or otherwise of the programme.

Business skills include cost-funding reconciliation reporting, HR management and leadership skills, budget information and control, including forecasting, payment processes, etc.

Strategy and planning

Project and programme management skills

1. Implementation and monitoring of adherence to programme management procedures

2. Sound technical knowledge in a variety of disciplines involved in the delivery of the projects and programme

3. Understanding of what is involved in identifying and developing potential projects from feasibility through evaluation, culminating in the production of a business case, including investment appraisal and identification and implementation of the most suitable procurement strategy for a portfolio of projects

4. Proven, successful experience in the management of alliance or partnering working, co-locating a team within a client organisation

5. Successfully coordinating activities including the ability to organise the workload of the team, balancing priorities and scheduling resources; able to deal with problems on own initiative and to make sound and timely decisions on a day-to-day basis

6. Ensuring the coordination and identification of programme risks and the management of risks through the development and maintenance of relevant tools

D.2 Programme business change manager

D.2.1 Main duties

The role of business change manager is mainly benefits-focused. The business change manager is responsible, on behalf of the programme sponsor, for defining the benefits, assessing progress towards realisation and achieving measured improvements. The business change manager role is associated mainly with programmes that tend to be more benefits-focused than projects, although projects that deliver benefits in their own right may warrant the creation of a business change manager role.

The role of business change manager must be the 'business side' in order to bridge the programme and business operations. Where the programme affects a wide range of business operations, more than one business change manager may be appointed, each with a specific area of the business to focus on.

The business change manager is responsible for the following:

1. Ensuring that the interests of the programme sponsor(s) are met by the programme

2. Obtaining assurance for the sponsoring group/programme sponsor that the delivery of the new capability is compatible with the realisation of benefits

3. Working with the programme manager to ensure that the work of the programme, including the scope of each project, covers the necessary aspects required to deliver the products or services that will lead to operational benefits

4. Working with the programme manager to identify projects that will contribute to realising benefits and achieving outcomes

5. Identifying, defining and tracking the benefits and outcomes required of the programme

6. Ensuring that maximum improvements are made in the existing and new business operations as groups of projects deliver their products into operational use

7. Leading the activities associated with benefits realisation and ensuring that continued accrual of benefits can be achieved and measured after the programme has been completed

8. Establishing and implementing the mechanisms by which benefits can be delivered and measured

9. Taking the lead on transition management, ensuring that business as usual is maintained during the transition and the changes are effectively integrated into the business

10. Preparing the affected business areas for the transition to new ways of working

11. Optimising the timing of the release of deliverables into business operations

D.2.2 Key competencies

The individual appointed as business change manager should be drawn from the relevant business areas, wherever practicable. Their participation in the programme should be an integral part of their normal responsibilities to enable changes resulting from the programme to be firmly embedded in the organisation.

Business change managers require detailed knowledge of the business environment and direct business experience. In particular, they need an understanding of the management structures, politics and culture of the organisation owning the programme. They need effective marketing and communication skills to sell the programme vision to staff at all levels of the business, and business change managers should ideally have some knowledge of relevant management and business change techniques such as business process modelling and re-engineering.

D.3 Programme benefits realisation manager

D.3.1 Main duties

The benefits realisation manager role is responsible for identifying, baselining, profiling, planning, tracking and reporting the benefits. The role involves developing and then managing the processes and management systems needed to support and govern effective benefits enablement and realisation to ensure the programme meets its objectives and realises its target financial savings.

The role is responsible for embedding and aligning the concept and principles of benefits realisation and contributes to a change in culture and behaviour across the programme in respect of benefits management and trains, educates and mentors, where appropriate, those staff directly involved in the delivery of business benefits.

1. Develops and supports the benefits management strategy and ensures that it reflects the direction of travel within the business and continues to be fit for purpose

2. Defines the benefits policies and procedures for the organisation

3. Defines, evaluates, recommends, monitors and assures benefits derived from component projects and the overarching change portfolio across the whole investment life cycle

4. Defines, manages and updates the organisation's benefits map against investment outcomes, profiles, interdependencies and realisation plans

5. Provides assurance that all selected component projects are aligned to the agreed benefits strategy and map and any impact identified

6. Provides the cost–benefit analysis data of the component projects' business cases and how these align to the portfolio benefits map

7. Supports management's decision-making by analysing benefits options and predicting future costs

8. Supports strategic business change by developing working practices that link benefits management into efficiency planning, performance measurement and 'value for money' delivery, ensuring benefit-led project prioritisation

9. Monitors benefits realisation plans and benefits review schedules

10. Ensures benefits owners are in place and the benefits are profiled, communicated, understood and managed

11. Analyses variances and initiates corrective actions with the benefits owners

12. Reviews the impact on the organisational benefits of new projects and change requests

13. Provides assurance to the organisation that the benefits are measurable, realistic and achievable and that the risks to the benefits are being effectively managed

14. Expedites delivery of benefits by establishing and maintaining working relationships with the sponsoring board, business change managers, project/programme managers and other key stakeholders to ensure the benefits are planned and realised

15. Initiates benefits reviews to provide assurance of benefits realisation plans

16. Is responsible for enhancing the organisational understanding and knowledge of benefits management

17. Maintains industry-standard professional and technical knowledge

18. Prepares reports by collecting, analysing, and summarising information and trends as requested by the programme management office or other performance/governance bodies

19. Monitors benefits from trends and analysis methods from other organisations

20. Attends relevant project and programme boards and departmental meetings to provide updates on benefits management and to provide practical advice to support delivery

21. Identifies proactively business benefits opportunities by liaising with key organisational stakeholders and assists, offers advice and guidance to enable a systematic business benefits process to be established

22. Monitors benefits realisation activity after component project delivery

23. Reviews benefits realisation achievements and puts continuous improvement processes in place

24. Identifies benefits within the various stages of the business case development

D.3.2 Key competencies

1. Have training in benefits realisation management with 5 years plus experience in a relevant field

2. Have recent financial accountancy experience

3. Demonstrate experience with the development of benefits management strategies, techniques, processes and tools

4. Demonstrate experience of cost–benefit analysis methods, benefits mapping and benefit-profiling tools

5. Have proven record of stakeholder engagement and working directly with executive teams, programme sponsors and corporate finance

6. Have recent experience in the development and implementation of management information processes and products related to benefits realisation

7. Able to apply structured business improvement techniques to identify business benefits

8. Have financial accountancy experience

9. Able to understand the strategic aims and objectives of the organisation

10. Demonstrate strong numerical and verbal critical reasoning ability

11. Demonstrate strong financial accountancy skills in terms of defining and projecting future benefits and associated costs

12. Able to analyse both qualitative and quantitative benefits information

13. Possess a high degree of accuracy and attention to detail

14. Demonstrate leadership of, and a positive approach to benefits management, demonstrating a willingness to challenge existing practices to support the organisation to continuously deliver benefits

15. Able to mentor and coach project managers and other practitioners in the benefits management processes

16. Demonstrate experience and competence in the use of MS Office applications (specifically Word, Excel and PowerPoint)

17. Demonstrate a personal commitment to own professional development

18. Able to recognise where processes are required and to develop and improve existing processes

D.4 Programme financial manager

D.4.1 Main duties

1. The core responsibility of this role is the coordination, control and reporting of cost information related to the programme

2. Client reporting requirements include working with project teams and finance to review delivery organisation information and then assemble individual

project reports, including quality checks; working with the programme manager to compile overall programme report and quality checks; ensuring reports meet customer requirements, including analysis of costs at project and pro-gramme level, variance analysis, forecasts, cost plans, budget requirements, third-party funding, etc., and implement agreed performance indicators to monitor projects more effectively; in addition, ensuring substantiation and audit trail are maintained

3. Supervise other commercial resources that may be supplied by the programme office to ensure reporting requirements for all projects within the programme are met

4. Interface with head of programme, programme managers and project man-agement teams, and manage cross-project dependencies from a high-level business perspective

5. Implement a structured programme management cost information gathering and reporting methodology, including supporting processes, procedures and tools

6. Promote the programme to all key stakeholders

7. Build and maintain strong relationships with senior colleagues within the client organisation

8. Support project management teams to ensure successful project delivery

9. Contribute with other programme managers for planning and control, cost and finance to ensure accuracy and uniformity of project reporting across the pro-gramme

10. Assist in supporting the interface with finance for affordability analysis across the programme and other corporate information exercises

11. Contribute to the programme risk and opportunity review by identifying possible conflicts and synergies visible through commercial analysis

12. Implement a structured programme management cost information gathering and reporting methodology, including supporting processes, procedures and tools

D.4.2 Key competencies

1. Significant commercial and cost management experience with proven track record of cost planning, monitoring and control, applying tools, principles, skills and practices to major programme of work

2. Knowledge of programme and project management methodologies and experi-ence in tailoring generic approaches to practical business situations

3. Effective interpersonal and communication skills, verbal as well as written

4. Able to find ways of solving or pre-empting problems and flexibility to be able to react to change in a positive manner

D.4.3 Key criteria

1. Estimating and/or reviewing capital cost estimates

2. Identifying and/or reviewing operating costs

3. Evaluating different funding sources and making recommendations

4. Managing cash flow (aligned with the project schedule)

5. Projecting revenue

6. Assessing risks/opportunities and associated costs

7. Evaluating procurement strategies and making recommendations

8. Identifying programme-level savings (supply chain, strategic purchases)

9. Assessing indexation/inflation

10. Managing and controlling cost including change control

11. Implementing common CBS (Cost Breakdown Structure)/WBS (Work Breakdown structure)/OBS (Organisation Breakdown Structure) required for the programme

12. Assessing trends, sampling, measuring, benchmarking, whole life costing

13. Providing business case analysis, project gateway reviews, and lessons learned

D.5 Head of programme management office

D.5.1 Main duties

1. Monitoring, independently reviewing, and reporting on the delivery of the programme

2. Establishing robust programme delivery reporting across the domain using the existing system, reports and tools available, or to set up new systems where necessary

3. Establishing independent health check criteria on programmes that will provide an independent view of delivery successes, risks and issues

4. Performing regular independent health check reviews on a material portion of the programme

5. Setting up and running high-level independent health check meetings

6. Updating the programme sponsor on overall programme delivery, identifying key delivery challenges, and proposing viable solutions to risks and issues in conjunction with programme managers

7. Ensuring programme static data and ongoing delivery updates are accurately captured in clarity, usable for the appropriate audience, and beneficial for future planning and prioritisation exercises

8. Managing the team of PMO resources

9. To act as a trusted partner and adviser around programme delivery

D.5.2 Key competencies

1. Have a technology background, including both programme/project management and application development experience

2. Have management experience working within a change division

3. Demonstrate an understanding of programme management and change practices

4. Show a desire to provide independent, agnostic oversight on a large portfolio of programmes separate from the teams actually owning and delivering the programmes

5. Have technical development experience, an essential but not primary focus of the role

D.5.3 Key criteria

1. Educated to degree level in technology or engineering from a university

2. Experienced in a financial services institution is desirable but not mandatory

3. Able and willing to manage and control detailed metrics, risks, and issues related to technology programme delivery

4. Experienced in using management information software applications is desirable

5. Experienced with corporate strategies, organisational structure and business policies and procedures in order to provide senior leadership

D.6 Programme risk manager

D.6.1 Main duties

1. Produce and manage the risk register

2. Own and promote the risk management process as defined in the quality management system

3. Produce and manage the risk register

4. Advise project teams on 'best practice' project risk and opportunity methodology

5. Conduct quantitative schedule risk analysis on the schedules

6. Conduct quantitative cost-risk analysis on the cost plans

7. Assist project teams to manage project risks as part of the individual project risk registers

8. Attend and participate in risk and opportunity management workshops where required by the project teams

D.6.2 Key competencies

Leading others

1. To promote risk and opportunity management throughout the programme and projects teams

2. To act as the point of contact for all technical/specific risk and opportunity management-related queries

3. To produce programme risk and opportunity management process proposals and communicate

4. To coordinate management of programme risk and opportunities register

5. Project and programme management

6. To ensure risk management methodology is fully incorporated within programme and project management

7. To offer advice and recommendations on risk plus opportunity management to project teams

8. To understand how risk management fits into the overall project and programme lifecycle

Analysis and use of evidence

1. To understand risk and opportunity products and processes to inform project and programme reports

2. To understand programme implications of project risks

D.6.3 Key criteria

1. Project risk management skills

2. Technical risk skills such as the use of risk management software, risk modelling skills and risk analysis skills

3. Understanding of project management lifecycle, planning and cost principles

4. Workshop facilitation skills

5. Presentation skills

6. Report-writing skills

7. Problem-solving skills

8. Communication

9. Process implementation management

D.7 Programme scheduling manager

D.7.1 Main duties

1. To set up, maintain and use the master programme schedule and key milestones from the individual projects analysed at programme level to identify cross-project critical paths, schedule risks, resource peaks/troughs, etc. This will involve detailed and regular communication with the project management teams to ensure that robust bottom-up information is flowing from the projects to enable programme-level analysis

2. To analyse potential problem areas – observations and recommendations for action will be required. Establish cost for cash flow and funding, commercial risks, cross-reference with risk for schedule risk analysis, etc.

3. To manage schedule control system (master schedule), conceived to serve as the management tool for planning, monitoring and controlling the design, procurement and construction of individual projects and overall programme at a strategic level

4. To achieve the programme goals through the development of a well-defined and realistic plan

5. To provide a visual means of conveying the information contained in the plan to stakeholders

6. To facilitate regular updating and monitoring of the programme

7. To prepare a master schedule that contains all relevant time schedule information from the individual projects

8. To ensure that the project teams update their current schedule with actual executed information and submit this each period. Review the submitted reports and compare them with the master schedule. Assessments sheets will be produced four times a week

9. To examine the period progress report from the delivery organisation project teams by the responsible project manager in conjunction with the relevant key personnel

10. To update the master schedule on a monthly basis. The master schedule will form part of the periodic report to the programme board

11. To undertake milestones trend analysis to identify all relevant project and programme milestones

12. To prepare a cost-loaded programme at an agreed-upon level to get qualified and schedule-interdependent information for the cash flow

13. To connect the information for risk to the master programme

14. To add to the general section of the master schedule to reflect key decision points and milestones

15. To continue communication with project managers to remind them of standard and frequency of time schedule reporting

16. Key competencies

17. A general understanding of the interfaces and interdependencies between the projects/departments

18. Qualities to lead people

19. Highly effective interpersonal and communication skills

20. Ability to find ways of solving or pre-empting problems

21. Flexibility to be able to react to change in a positive manner; willingness to provide support in areas outside core role for the overall benefit of the programme management team

D.7.2 Key criteria

1. Significant schedule management experience with proven track record of activity planning, monitoring and control applying tools, principles, skills and practices to major programme of work

2. Knowledge and experience in the established tools from Microsoft and Primavera software

3. Knowledge of programme and project management methodologies and experience of tailoring generic approaches to practical business situations

D.8 Programme cost manager

D.8.1 Main duties

1. Check/validate applications for payment from delivery organisations with reference to numerical accuracy, allowable/disallowable costs, duplication, valuation against progress and valuation against elements of project

2. Advise the project manager and assist with resolution of anomalies

3. Assist finance with checking/validation of invoices from delivery organisations

4. Review and confirm (as deemed appropriate) delivery organisations' estimated costs, operational expenditure, revenues and forecasts

5. Work with delivery organisations to increase levels of confidence in financial information being provided both on a periodic, annual and out-turn basis

6. Act as 'bridge' to facilitate better reporting of commercial issues and their subsequent financial impact

7. Review (and align) delivery organisations' accruals methods

8. Facilitate early intervention through speedy identification of issues affecting projects with major implications for project financing

9. Identify early any issues relevant to current funding/budget availability and help provide clarity on impact of issues such as rollovers or transfers on budget

10. Deliver project cost information that is accurate, timely and reliable

11. Ensure tight commercial, financial and business controls are in place

12. Measure project performance against objectives, forecasts and budgets

13. Assist in the clear presentation of project expenditure

14. Prepare and analyse periodic cost reports at the programme level

15. Identify process improvement opportunities

D.8.2 Key competencies

1. Understand applications for payment and invoice process – to liaise with delivery organisations and project teams to ensure TS pays only for valid services or products and to ensure value for money

2. Have experience in management accounting, including project reporting – to review and interrogate delivery organisations' reports to TS to ensure accuracy, consistency and completeness

3. Understand capital grant funding, RAB financing and other available funding options and the associated reporting requirements – to enable preparation of the TS affordability models

4. Interface between financial and project/programme teams

D.8.3 Key criteria

1. Have experience preparing project accounting information

2. Have experience with contracting or project environment

3. Be commercially aware, with strong analytical and communication skills

4. Be adaptable and a motivated self-starter

5. Have strong interpersonal skills in non-financial management

6. Demonstrate planning, organisational and analytical skills

7. Have a high degree of computer literacy, including spreadsheet and MS Office skills

Bibliography

Budzier, F. (2018). *Quantitative Cost and Schedule Risk Analysis of Nuclear Waste Storage. Oxford Global Projects.*

Flyvbjerg, B. (2004). *Procedures for Dealing with Optimism Bias in Transport Planning. The British Department for Transport.*

Kahneman, T. (1977). *Intuitive Prediction: Biases and Corrective Procedures. Cybernectics Technology Office, Defense Advanced Research Projects Agency.*

Code of Practice for Programme Management in the Built Environment, Second Edition. The Chartered Institute of Building.
© 2024 John Wiley & Sons Ltd. Published 2024 by John Wiley & Sons Ltd.

Past Working Group of the *Code of Practice for Programme Management*

First Working Group for the Code of Practice for Programme Management

Saleem Akram BEng (Civil) MSc (CM) PE FIE FAPM FIoD EurBE FCIOB	Director, Construction Innovation and Development, CIOB
Gildas André MBA MSc BSc (Hons) MAPM MCIOB	Managing Partner, GAN Advisory Services
David Haimes BSc (Hons) MSc MCIOB	Strategic Programme Director, Manchester Airports Group
Dr Tahir Hanif PhD MSc FCIOB FAPM FACostE FIC CMC FRICS	Project Control Specialist, Public Works Authority, (Ashghal), State of Qatar
Stan Hardwick FCIOB EurBE	Global Contracts Manager – Procurement, Specsavers
Dr Chung-Chin Kao ICIOB	Head of Innovation & Research, CIOB
Arnab Mukherjee BEng(Hons) MSc (CM) MBA FAPM FCIOB	WG Technical Editor
Andrew McSmythurs BSc FRICS MAPM	Director of Project Management at Sweett (UK) Ltd and RICS Representative
Paul Nash MSc FCIOB	Director, Turner & Townsend, Senior Vice President, CIOB
Piotr Nowak MSc Eng ICIOB	WG Secretary, Development Manager, CIOB
Dave Phillips FAPM CEng	Divisional Director, Mott MacDonald
Milan Radosavljevic PhD UDIG MIZS-CEng	University of the West of Scotland
Dr Paul Sayer	Publisher, John Wiley & Sons Ltd, Oxford
Roger Waterhouse MSc FRICS FCIOB FAPM	WG Chair, University College of Estate Management, Royal Institution of Chartered Surveyors, Association for Project Management
David Woolven MSc FCIOB	WG Vice Chair/Editor – University College London

Past Working Group

The following also contributed in development of the *Code of Practice for Programme Management*.

Susan Brown FCIOB MRICS	Property Asset Manager at City of Edinburgh Council
Jay Doshi	ICE Management Panel Member (Director, Amey Ventures)
Nikki Elgood	WG Administrator, CIOB
Una Mair	WG Administrator, CIOB
Simon Mathews	Director/HLG Associates
Gavin Maxwell-Hart BSc CEng FICE FIHT MCIArb FCIOB	Head of Contract Management, AREVA CIOB Trustee, Non-Executive Director, Systech International
David Merefield	Head of Sustainability, Property, Sainsbury's Supermarket Ltd
Alan Midgley	Medium Risk Reviewer, Cabinet Office Director, AGMidgley Ltd
Dr Sarah Peace BA (Hons) MSc PhD	Consultant
David Philp MSc BSc FICE FRICS FCIOB FCInstES FGBC	Global BIM/MIC Director – AECOM, RICS Certified BIM Manager, CIOB Ambassador
Tony Turton MBA FICE	Project Development and Production Director, Highways England

Index

Note: Page numbers in *italic* refers to *figures*.

Code of Practice for Programme Management in the Built Environment, Second Edition. The Chartered Institute of Building.
© 2024 John Wiley & Sons Ltd. Published 2024 by John Wiley & Sons Ltd.